East Mediterranean Natural Gas: A Game Changer?

East Mediterranean Natural Gas: A Game Changer?

Basel Nashat Asmar

2050 consulting

First published in the United Kingdom in 2024 by
2050 Consulting
48 Imperial Hall
104-122 City Road
London EC1V 2NR
United Kingdom

A CIP for this book is available from the British Library.

ISBN 978-0-9567368-3-3

Author's Notes:
1. The material in the book made no use of any proprietary data owned by S&P Global and does not express S&P Global's opinions. The opinions expressed in this book are the responsibility of the author.

2. The use of particular designations of countries or territories does not imply any judgement by the author as to the legal status of such countries or territories, of their authorities and institutions or of the delimitation of their boundaries.

Printed in the United Kingdom

Cover design by Eman Faidi

This book is dedicated to the memory of my aunt who devoted her life to establish a just and prosperous Eastern Mediterranean Region.

LIST OF CONTENTS

ACKNOWLEDGMENTS

I would like to thank my esteemed expert friends and colleagues for their insightful comments on the draft chapters for discussing and challenging key ideas in the manuscript. Special mention goes to Andrew Day and Paul Markwell who tested some of the narrative and helped shape the book's direction, to Nick Thomson and Prashant Pillai for their insights, and to Pritesh Patel for his support, encouragement and critical comments.

I am deeply grateful to my family—my beloved mother, brothers, sisters-in-law, and nephews, Sanad and Qais—as well as my friends in the UK, Jordan, the USA, and the UAE, for their unwavering encouragement and support throughout the long writing process. Special thanks to Ehab Wazani for sharing his sports expertise and to Fareed Bilbeisi for his invaluable observations.

Once again, completion of this book would have not been possible without the efforts of Karen Hall, who not only edited my manuscript meticulously, but also challenged my thinking by raising shrewd questions and sharing valuable insights which influenced the final book shape.

ABOUT THE AUTHOR

Basel Nashat Asmar is an expert in oil and gas fundamentals, costs and technology, a dynamic simulation expert with extensive computer, modelling and simulation skills. He is currently a Director with S&P Global (acquired IHS Markit in 2022), based in London, UK. Dr Asmar worked previously in several roles with major engineering companies, concerned with large liquefied natural gas (LNG) regasification and liquefaction terminals, natural gas compression stations, and offshore oil and natural gas production platforms. He worked as a senior process engineer and dynamic simulation specialist with CB&I (currently McDermott), Mott MacDonald, and IMEG. He was a lead consultant with 2050 Consulting Ltd and Trident Consultants Ltd. Prior to that, he worked in academia as a research associate at the University of Nottingham.

Dr Asmar has authored three books and published more than 50 articles in specialist international journals, for conference proceedings and newspapers. He is a Chartered Engineer, a Member of the Institution of Chemical Engineers (IChemE), a Senior Member of the American Institute of Chemical Engineers (AIChE) and a Member of the Jordanian Engineers Association. He holds a BSc in Chemical Engineering from the University of Jordan, an MSc in Process and Project Engineering, a PhD in Chemical Engineering from the University of Nottingham, and a doctorate in Geoscience from Freie Universität Berlin, Germany.

FOREWORD

By
Karim Kawar
President of Kawar Group[1,2]
Amman, Jordan

I am pleased to be invited by Dr Asmar to write the foreword for this engaging book. The subject of natural gas and renewables is critical in today's world, here Dr Asmar's thorough exploration of the situation in the Eastern Mediterranean, puts its role into context. While the book includes a lot of data, which can make understanding the connections difficult, he presents it in such a way that it remains accessible to those of us who are not experts but still have an interest in the subject.

The Eastern Mediterranean region holds significant natural gas reserves, but they are not that impressive when compared to major players like the USA, Qatar or Russia. Currently, these reserves represent about 2.6% of the global total or 1.9%, when excluding Libya. In contrast, Qatar holds 11.2% and Iran has 16%. Even with potential new discoveries, the region's position in the global gas market is modest at best. The production levels are also low, with most of the gas consumed domestically. Only Israel and Libya are exporting notable quantities and, even then, they account for just around 1.5% of global exports. Additionally, Egypt's LNG re-exports complicate matters, as they represent a form of circular trade rather than new production.

---

[1] The author is expressing a personal point of view here. The opinions stated in his preface should not be regarded as the official position of Kawar Group.
[2] Kawar Group, founded in 1926 and based in Amman, Jordan is a privately-owned group of companies serving the MENA region, with diversified interests in information and communication technology, energy, cleantech, shipping and transportation, travel and tourism, insurance, healthcare, international trade and development.

In the regional power dynamics, Israel has become a key player, using its natural gas finds to boost its economy and geopolitical influence. Egypt, while well-placed geographically and possessing the necessary infrastructure, faces internal struggles, declining gas production and economic issues. Turkey has a long-term strategy, already serving as an energy hub and its influence will likely grow as it develops its own Black Sea gas reserves.

The region is a critical transit point for global energy trade, particularly for pipelines and LNG shipments. While it plays a vital role in facilitating energy from other regions, it does not supply large quantities of gas itself. This means that the natural gas sector in the Eastern Mediterranean is a small but strategically important part of the global energy landscape.

Addressing the idea of energy-related conflicts, the book argues that historical issues in the region have generally been driven by political and ideological disputes rather than energy resources. The ongoing Gaza War, for example, is primarily about territorial and ideological differences, not energy. However, water scarcity is emerging as a more pressing issue that could lead to future conflicts.

In summary, the Eastern Mediterranean's natural gas situation is not a game changer but also not insignificant. While its reserves and production are minor in the global context, its strategic location and political influence give it a notable role in regional energy dynamics. There are time-sensitive opportunities, particularly in LNG development, that could increase its importance, but delays and political instability may hinder its potential. Ultimately, the region's role in natural gas is more of a niche player in the broader energy transition.

The author uses a sports analogy to explain the geopolitical situation in the Eastern Mediterranean. It aims to simplify complex natural gas policies for non-experts. Through this analogy, the book highlights the chaotic nature of the geopolitical landscape in the Eastern Mediterranean, where self-interested actions lead to disarray. While the analogy provides a clearer picture, it also emphasises that the region's natural gas resources, though not significant on a global scale, could still make a difference in the lives of its people and contribute positively to the energy landscape if the right conditions are met.

This book is a valuable resource for anyone looking to understand the complexities of natural gas situation in the Eastern Mediterranean. It sheds light on the region's geopolitical dynamics, making the information accessible to a broader audience, who might be unfamiliar with it. I encourage readers to delve into its pages, explore the intricate connections between natural gas, politics and the future of energy in this vital part of the world.

Karim Kawar
Amman
October 2024

INTRODUCTION

The potential of East Mediterranean natural gas has been a big story in the Middle East, lately getting significant coverage globally in the industry and specialised international media. The idea of writing this book started after noticing a number of Middle Eastern media news cycles that gave often misleading information about the potential of the East Mediterranean natural gas and its potential to be a game changer in the global natural gas market. These stories often cited the natural gas reserves and production as important reasons for the political strategies of the region's governments, some of their backers, as well as their adversaries.

The Russian invasion of Ukraine and its impact on natural gas prices, along with the Armageddon headlines painting a picture of frozen Europe that needs saving, in the winter of 2022/23, stimulated people's interest in learning more about natural gas and where it is sourced from. As a result, this placed additional emphasis on the status of Eastern Mediterranean natural gas. Obviously, we know now, after more than two years passed, that this Armageddon was a myth, and that Europe did not descend into a war generated 'Frozen Winter'.

In this book I take a simplified approach to tackling the story of this East Mediterranean natural gas resource. I look at it from several angles, examining the factual data. I also attempt to debunk the numerous exaggerated claims and conspiracy theories that place East Mediterranean natural gas in the middle of a narrative claiming to drive peace and security in the Middle East. Finally, I attempt to answer a fundamental question – Is East Mediterranean natural gas really a regional and global game changer?

It is my intention to focus on the natural gas story, therefore any analysis of oil markets in the region is outside the scope of this book and is not discussed in it.

In my opinion, a major geopolitical game is happening in the region, that uses natural gas policies as a proxy. Thus, in this book, I refer to the various stakeholders as 'players', each of whom has his place to be positioned and his role to perform.

The Eastern Mediterranean's natural gas reserves, while often hyped, are actually relatively small in the global context. Combined reserves from the region account for about 1.9% to 2.6% of global natural gas reserves,[1] which pales in comparison to neighbouring giants Qatar (11.2%) and Iran (16%). Moreover, despite its geopolitical significance, the region's natural gas production is mainly consumed domestically, with minimal exports, so its contribution to the global natural gas market is actually quite minor. Most exports from the region are internal, with Israel and Libya being the only net natural gas exporters. However, future prospects do exist for additional exports, most probably through LNG, although political instability and delays in project development continue to hinder progress.

Currently Egypt, Greece, and Turkey are vying for dominance and competing to become 'The' energy hub. Egypt benefits from existing infrastructure but faces internal economic and political challenges, as well as natural gas deficit due to growing domestic demand. Turkey, on the other hand, is showing the most promise with numerous pipelines connecting it to global markets and is benefiting from support from Russia to strengthen its position as a natural gas hub, while Greece's reliance on Turkey for energy strategy, creates vulnerabilities and threatens Greek security. In the short term, with Libya still engulfed in political upheaval, Israel stands out as the region's main exporter, though ongoing conflicts may affect future prospects.

This book is arranged in five chapters. It starts by giving a brief introduction on natural gas and its role in the ongoing energy transition. Then it analyses the Eastern Mediterranean Region, focusing on political and hydrocarbon related issues. It then

---

[1] Depends on if Libya is included or excluded.

follows with a detailed analysis of the natural gas status in each of the countries (players) in the region along with some other countries that also influence the narrative. The concluding chapter brings the story together, joining all the dots and putting it all in perspective. Since this book is not intended to be technical, interested readers can access essential technical definitions and data in the appendices, if they desire.

This manuscript started few years back, but stuttered as the rapid developments in the region meant it had to keep being rewritten, with rapid changes forcing me to revise again and again. The ongoing Israel-Gaza War of 2023-present, and its expansion into other countries in the Middle East, will undoubtedly also affect the future narrative in the region and with it, its natural gas story.

Chapter 1
NATURAL GAS AND THE ENERGY TRANSITION

Natural gas is a fuel that forms a crucial part of the energy jigsaw picture in the world. With the momentum of energy transition accelerating, and the loud voices made by the 'woke-like' green lobbyists and media, the issue of natural gas has become a controversial one. Some consider it to be part of the process of decarbonisation, an essential component of and bridge in the energy transition journey. Others are firmly against it and oppose its continuing usage, since it is one of the main sources of emission from fossil fuels that they argue must be eliminated. In this chapter, I present a crash course on natural gas for non-experts, therefore, knowledgeable energy-literate readers can skip this chapter if they prefer.

1.1   Natural Gas 101

Unlike solid and liquid substances, natural gas is a gaseous, colourless, odourless fossil fuel[1] that consists of a mixture of hydrocarbon compounds and a small amount of various non-hydrocarbon components. The prime hydrocarbon compound is methane. Other hydrocarbon compounds that can be found in significant quantities are ethane, propane, butane and pentane. Heavier hydrocarbons can also be found in small quantities. Non-hydrocarbons that found include primarily carbon dioxide, nitrogen, hydrogen sulphide, helium and trace metals such as mercury. The exact composition of natural gas varies widely between reservoirs. Table 1.1 outlines natural gas typical makeup.

---

[1] Fossil fuel is defined as hydrocarbon-containing material such as coal, oil, and natural gas, formed naturally in the Earth's crust, from the remains of dead plants and animals, that is extracted and used as a fuel.

Table 1.1: Typical composition of natural gas

Component	Range %	
	Non-Associated Natural Gas	Associated Natural Gas
Methane	70 - 98	45 - 92
Ethane	1 - 10	4 - 21
Propane	trace - 5	1 - 15
Butanes	trace - 2	0.5 - 7
Pentanes	trace - 1	trace - 3
Hexanes	trace - 0.5	trace - 2
Heptanes and heavier	trace - 0.5	0 - 1.5
Nitrogen	trace - 15	trace - 10
Carbon Dioxide	trace - 5	trace - 4
Hydrogen Sulphide	trace - 3	0 - 6
Helium	0 - 5	0

Source: Standard Book of Petroleum and Natural Gas Engineering, W C Lyons and G J Plisga, Gulf Professional Publishing, 2005.
Note 1: Some natural gas fields have majority Carbon Dioxide or Hydrogen Sulphide.

At natural reservoir conditions, natural gas can exist in gaseous, liquid or even solid form. In gaseous form, natural gas is found isolated in natural gas and condensate fields; it occurs as a cap above oil fields in oil reservoirs; and it can be found underground in rock, coal, or porous sand formations. As liquid, natural gas is found dissolved, forming a solution with oil in underground oil fields as well as under extreme pressure and temperature conditions. As a solid, natural gas is found trapped in crystals in the form of Methane Hydrates.

When natural gas is produced from a reservoir that does not contain significant quantities of crude oil, it is termed *non-associated* gas. However, when it is produced, often as a by-product, with crude oil from a reservoir, it is termed *associated gas*. The latter term is applicable whether the gas is found dissolved in the oil or if it is free gas, above the oil, in the reservoir.

Natural gas is commonly extracted by drilling from wells onshore or offshore. Since natural gas is lighter than air, the gas, which is naturally pressurised, will rise to the surface with little or no interference. This is usually the case in gas and condensate wells. However, extracting associated gas from oil wells is different, requiring lifting processes such as pumping (i.e. pumping the oil in which gas is dissolved), prior to separating it.

Several methods can be applied to optimise the recovery of natural gas. Primary recovery methods extract natural gas to the surface utilising the reservoir's underground pressure. When this pressure is depleted, enhanced gas recovery methods (EGR) are employed to extract more natural gas from the reservoir. This is done by injecting water or other gases, such as carbon dioxide, to repressurise the reservoir. Additionally, stimulation techniques such as hydraulic fracturing, acidizing, and dewatering, are used to extract natural gas from 'unconventional' resources such as tight natural gas, shale gas and from coal beds. Natural gas can also be extracted via chemical transformation of oil or coal.

It is important to state that, from an economical perspective, there is a fundamental difference to consider when producing associated or non-associated natural gas. In the case of the associated natural gas, the main objective of developing the reservoir is to produce and sell crude oil, so the associated natural gas may be wasted (by being flared, vented; or injected) or collected then utilised as by-product and sold to consumer. The case of non-associated gas is different as the reservoirs will only be developed if there is a sales route determined to the produced natural gas.

Natural gas extraction methods and technologies is a vast subject, so they are outside the scope of this book. Interested readers can find a wealth of knowledge elsewhere.[2]

1.1.1 Natural Gas Classifications
Natural gas is classified according to several criteria, which are not standardised. These standards are currently inconsistent and, confusingly, differ between countries and organisations. However, unlike oil, which is classified based on physical characteristics such as density and viscosity, natural gas classification does not take these characteristics into consideration as it has almost

---

[2] Standard Book of Petroleum and Natural Gas Engineering, W C Lyons, G J Plisga and M D Lorenz, 3rd Edition, Elsevier, 2016

uniform physical characteristics, thus differing only slightly in terms of density and viscosity.

In terms of chemical characteristics, natural gas is classified based on three different criteria:

- Hydrocarbon composition - it is labelled *dry gas* if it is composed from almost pure methane, and *wet gas* if other heavier hydrocarbons are present in the gas mixture. *Wet gas* usually originates in condensate fields which produce raw natural gas as well as natural gas liquids (NGLs).
- Sulphur content - it is labelled *sour gas* if the gas contains hydrogen sulphide or other sulphur impurities, and *sweet gas* if the sulphur content is considerably low.
- Carbon dioxide content - it is labelled *acid gas* if the gas contains carbon dioxide. The same label applies if the gas contains sulphur dioxide, hydrogen sulphide or similar contaminants.

In terms of technical, economic, and geographical criteria, natural gas is classified as either *conventional* or *unconventional* (also referred to as *non-conventional*). In the broadest sense, conventional natural gas is the gas extracted from economically feasible deposits using fully developed practical and easy-to-implement methods. Unconventional natural gas is more difficult, and less economically feasible to extract. This is often because of the location or the nature of the deposit, or that the technology required to reach it has not been developed fully, or it is simply too expensive. Unconventional natural gas has the same chemical and physical composition as conventional natural gas, the 'unconventional' label is usually given because of the physical condition of the reservoir that the gas is found in.

With the advance in technology and geological knowledge, unconventional natural gas deposits are not only being reclassified as conventional deposits, but they are also making up an increasingly large percentage of natural gas reserves and their subsequent production volumes. As a result, the criteria for what is considered unconventional natural gas, is changing over time.

The economics of extraction plays a significant role in determining whether a particular deposit is unconventional or simply too expensive to extract.

In recent years, the classification of deposits of deep natural gas (typically deeper than 5000 metres onshore, or located in deep water fields offshore), once considered unconventional, has shifted, and they are now considered conventional. This is because of improved deep drilling, exploration, and extraction techniques, resulting in making the production of deep natural gas more economical, albeit still relatively more expensive than cheaper conventional natural gas. The same applies to natural gas deposits in the Arctic region, which were considered unconventional, not because of technical issues, but rather their geographical location. These deposits have also made the leap and moved from being considered unconventional to conventional deposits, as the rapid increase in the price of natural gas placed them firmly within the economic feasibility window. This profitability outweighed the logistical difficulties that previously hindered their extraction, and they are now being exploited in Russia and North America.

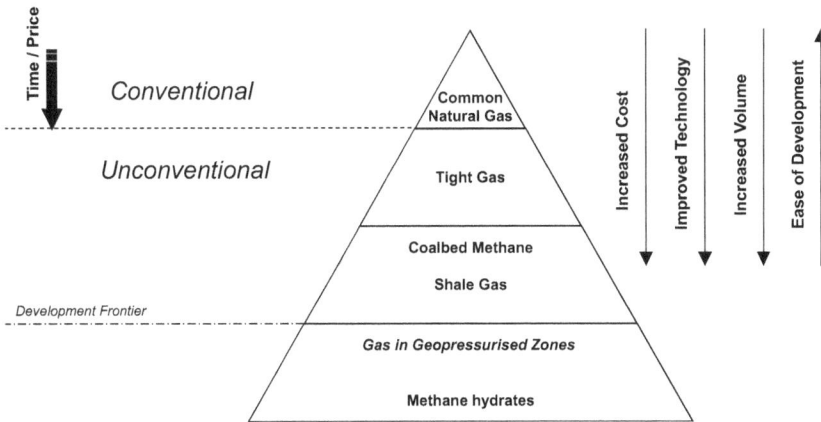

Figure 1.1: Natural gas resource triangle

Even though there is no universally agreed definition of

unconventional natural gas, broadly speaking, there are five main categories (although some experts may add or remove a category). In this book the five categories used are tight gas, shale gas, coalbed methane, gas in geopressurised zones and methane hydrates. Figure 1.1 presents the natural gas resource triangle which shows the natural gas categories and illustrates the relationship between cost, technology, and resource volume.

In simple terms, the easiest-to-extract natural gas (i.e. conventional natural gas) occupies the smaller upper parts of the triangle. It exists in smaller volumes, its deposits are easier to develop, plus they are more feasible and profitable with current technology. More difficult deposits (i.e. tight gas, coalbed methane and shale gas) are located lower on the triangle. They exist in larger volumes, but they are more difficult and expensive to develop, and thus, they are less profitable. At the bottom part of the triangle are the most difficult and speculative deposits (i.e. natural gas in geopressurised zones and methane hydrates). Although they exist in even higher volumes, they are prohibitively expensive to extract, they may have an unfavourable environmental footprint, or because there is no current feasible technology to extract them. The natural gas development frontier is defined by the current status of technically feasible development. It is influenced by the price of natural gas. Thus, the higher the price, the further down the triangle the frontier can move. The increase in price will overcome many obstacles by offsetting additional costs, encouraging more complex technology development and adaptation. Moreover, in the current atmosphere, where the momentum of energy transition continues to advance, concern for environmental aspects plays an increasingly stronger role in determining the potential future of natural gas deposits and their development.

1.1.2 LPG Versus Natural Gas
In the Eastern Mediterranean region, liquified petroleum gas (LPG) is colloquially referred to as 'gas' and is often confused by people with 'natural gas', where the word gas is used interchangeably for both. The confusion often extends to the media.

In fact, LPG is an oil product extracted by refining crude oil or wet natural gas. It consists of propane, butane, or a mixture of propane, butane and other relatively light hydrocarbons. It is often used in the Middle East for cooking and heating, although other uses in transport and industry exist. This is discussed in my earlier book.

1.2 Natural Gas Reserves and Production

Most industry experts, media outlets, and major data sources agree that the world's natural gas conventional reserves are continuing to increase. This upward trend is evident in Figure 1.2, which shows the conventional natural gas reserves since 1980 reported by the EIA, BP/Energy Institute (EI), and OPEC, with all sources reporting a substantial increase in proved reserves in excess of 140% between 1980 and 2023.

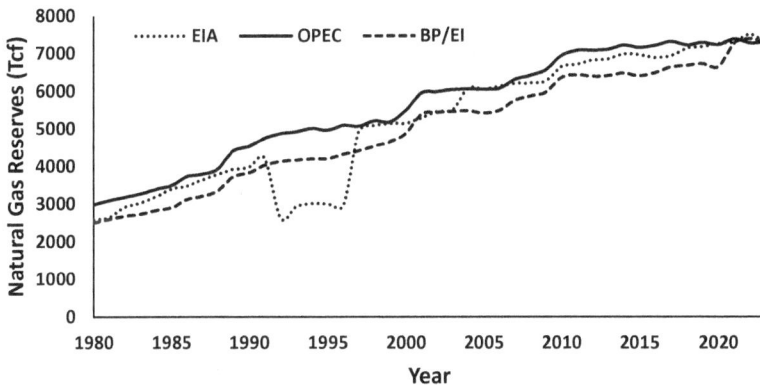

Figure 1.2: World natural gas reserves (1980-2023)

Source: EIA (http://www.eia.doe.gov/international); OPEC (Annual Statistical Bulletins 1999-2024); BP (Statistical Review of World Energy 2001-2022), EI (Statistical Review of World Energy 2023-2024).

This is not because of some geological miracle where natural gas is being produced constantly by the planet. Rather it is because of better exploration technologies and the way data is collected and interpreted. The data in Figure 1.2 used to exclude unconventional natural gas in the past, though this has changed considerably in the last decade. More recently, first the USA, then other countries increasingly began including substantial amounts of their unconventional natural gas reserves into their reported conventional reserves. For example, the EIA numbers published for the USA for 2023, include a large proportion of unconventional natural gas, encompassing unconventional plays of tight gas, shale gas, and coalbed methane, with many being reclassified as conventional reserves. The same has been done in many countries, including Canada and Australia. All the major data sources estimates are in broad agreement volumetrically, with small unavoidable differences being due to different practices in data reporting.

According to the Oil & Gas Journal at the end of 2023, the world's total proved natural gas reserves stood at 7506.576 Tcf. The leading three countries, in terms of natural gas reserves, are Russia, Iran and Qatar, accounting for 22.5%, 16% and 11.2% of total reserves respectively.[3]

The terminology of natural gas production can be confusing as different data sources often report different quantity types, which many media outlets mistakenly use interchangeably. There are three types of natural gas production terms used and it is important that they are differentiated: gross, marketed and dry production. *Gross production* is the full well stream produced from a natural gas well, excluding lease condensates. It includes natural gas plant liquids and all non-hydrocarbon gases. Parts of the gross production are reinjected into the natural gas wells to maintain their pressure, or they can be vented or flared. However, the majority of the gas, which is the remainder gas, after removing most non-hydrocarbon gases in treating or processing operations, is labelled *marketed production*. Finally *dry production* is the marketed production, less gas extraction loss, which includes gas lost in transmission and shrinkage. The shrinkage is the volume of natural gas that is transformed into liquid products during processing, primarily at natural gas processing plants.

[3] https://www.ogj.com/general-interest/economics-markets/article/14302481/global-oil-and-natural-gas-reserves-both-increase

Despite all the current noises made regarding energy transition, to date, the upward trend in natural gas production is all but certain and it is expected to continue as long as there is demand. This demand is projected to grow in the short term and medium term by all major forecasts. Unfortunately, these forecasts disagree on the long-term projection and if, or when, natural gas production will peak. In 2030, the demand is projected to have reached 158.9 Tcf and 151.8 Tcf by the EIA and IEA respectively, in their reference or stated policies cases. [4,5]

It is important to note that, historically, despite a few temporary blips, e.g. the 2008 financial crisis and Covid-19 pandemic, the global natural gas production capacity, or actual output, generally showed continuous growth. Any reported short intermittent declines, only affecting actual production, were due to localised domestic economic or political factors and technical or environmental incidents.

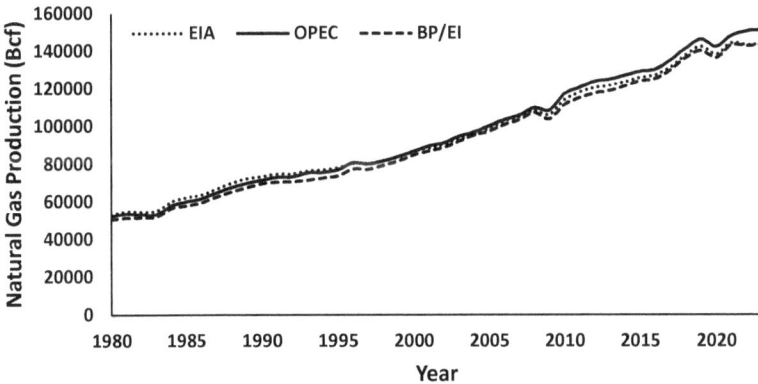

Figure 1.3: World natural gas production (1980-2023)

Source: EIA (http://www.eia.doe.gov/international); OPEC (Annual Statistical Bulletins 1999-2024); BP (Statistical Review of World Energy 2001-2022), EI (Statistical Review of World Energy 2023-2024).

Thus, this indisputable upward trend, to date, supported by all major data sources, indicates that the world's natural gas

---

[4] EIA, International Energy Outlook 2023
[5] IEA, World Energy Outlook 2023

production is continuing to increase. Figure 1.3 demonstrates this showing the reported total natural gas production since 1980, as published by the EIA, BP/EI, and OPEC, with all sources reporting an increase of over 180% in total natural gas production between 1980 and 2023. Total dry natural gas production numbers include both conventional and unconventional natural gas, but exclude NGL production. That is included within oil or liquid production, but they also exclude other gaseous fuels, including biofuels. It can be seen that the differences in the reported data are insignificant and that the major data sources not only agree on the upward trend, but also on the quantitative natural gas production. As expected, there are slight differences, but these differences can be attributed to different practices in data reporting (i.e. what exactly is included), with OPEC for example reporting marketed production, while the EIA reporting dry production.

Following the seismic hit the energy markets suffered in 2020, there was debate, not about whether fossil fuels demand will ever recover to its 2019 levels, but rather the timing of peak demand.[6,7]. In the case of natural gas, forecasts agreed that demand will continue to grow in the short term, although some suggest that peak natural gas demand, will occur only a few years after peak oil demand.[8] Opinions published by consultants (e.g. IHS Markit (currently S&P Global), Wood Mackenzie),[9] think tanks and organisations differ, with all agreeing that the Covid-19 pandemic has brought the peak oil time closer.[10]

However, three years on, natural gas markets proved to be more resilient. Despite another major shock the market endured in 2022 following the Russian invasion of Ukraine, natural gas demand continues to increase in the short-term. Dates of potential peak demand are being revised repeatedly, with some, such as IEA,

---

[6] https://www.petroleum-economist.com/articles/midstream-downstream/lng/2020/gas-and-lng-brace-for-tougher-times
[7] https://www.worldoil.com//news/2020/6/10/decline-in-natural-gas-demand-may-take-years-to-reset
[8] https://www.worldoil.com/news/2021/1/14/mckinsey-projects-a-2029-oil-demand-peak-accelerated-by-covid-19
[9] https://oilprice.com/ 15/05/2020
[10] https://www.worldoil.com/news/2021/1/14/mckinsey-projects-a-2029-oil-demand-peak-accelerated-by-covid-19

setting peak dates in 2030, while others, such as OPEC, not envisaging any peak in the near future.[11]

1.2.1 Natural Gas Transportation

Natural gas transportation requires customised infrastructure, which is one of the main hurdles affecting its universal use. This is mainly due to its low density, which imposes restrictions on the options for means of transport. Natural gas needs to be transported to consumers either after being processed onsite on production fields or after being moved from production to processing terminals nearer to consumers.

Three main modes of transport are in use:

- Pipelines - these are considered the most efficient and economical method to transport natural gas. Pipelines are built both over land (onshore) and on the seabed (offshore). They connect offshore facilities to onshore terminals, they also connect countries to each other and they extend for thousands of kilometres. Pipelines are used to move natural gas from the wellheads to gathering and processing facilities, from there to terminals and subsequently to distribution networks/grids. If economically feasible, they can also be used to supply major industrial sites and power stations directly. Natural gas pipelines range from distribution pipelines, (connecting consumers) to longer pipelines connecting cities, countries and even continents. They include numerous compression stations to maintain the gas pressure and facilitate its movement. Natural gas pipelines have both economic benefits and strategic value, that involves international security, therefore interlinking them with international politics. Disputes over natural gas pipelines management are often in the news. They are known to have contributed to international political crises, currently being used by exporting countries (e.g. Russia) or transit countries (e.g. Ukraine) as tools to pressurise and

---

[11] https://agsiw.org/opec-and-iea-at-odds-over-long-term-oil-demand-outlook/#:~:text=%E2%80%9CGlobal%20natural%20gas%20use%20has,for%20natural%20gas%20will%20peak.

influence the policies of the importing countries. As they are so significant, this makes them vulnerable as legitimate targets in military disputes as well as by other actors for vandalism, sabotage, and terrorists. The most famous example is the 2022 sabotage of the Russian Nord Stream pipeline (connecting Russia and Germany) in the Baltic Sea.[12]

- Ship carriers or tankers – which mostly transport liquefied natural gas (LNG), but a few also transport compressed natural gas (CNG) across water. Transporting LNG requires specialised facilities for liquefaction at the production point and regasification at end-use facilities or into a pipeline. As a result, ship carriers have a higher cost in comparison to oil tankers, which simply carry liquid at atmospheric temperature and pressure. However, using this mode of transport allows consumers access to global natural gas supply and thus reduces their dependency on single supplier. An additional benefit which also improves security of supply.

- Tank trucks - these can carry LNG or CNG on land over shorter distances. They may transport natural gas directly to end-users, or to distribution points i.e. to pipelines for further transport. This mode of transport is rarely used for international natural gas transport due to its considerably higher costs, compared to LNG carriers and pipelines.

There have recently been suggestions that natural gas could be transported in solid form as hydrates, however, that technology is still in its infancy, and is not yet feasible.[13]

For decades, most natural gas was transported globally using pipelines, though with the growth of LNG as a medium of natural gas trade, sea tankers are increasingly being used. At the moment road transport is used between Mexico and USA (which forms a

---

[12] https://en.wikipedia.org/wiki/2022_Nord_Stream_pipeline_sabotage
[13] https://www.linkedin.com/pulse/natural-gas-hydrate-alternative-optimistic-way-iit-madras/

tiny proportion of preferred methods).

Natural trade, via pipelines, accounted for over 41% of the world's total trade in 2023[14] Pipelines are the cheapest means of transporting natural gas, compared to using tankers for shorter distances. They have a distinct cost advantage for transporting gas up to 2000 km offshore and 4000 km onshore,[15] though greater distances are possible by increasing pipeline diameter and capacity. This economic advantage diminishes when increases in the transport distance are required but, after a certain threshold, tankers transporting natural gas have a cost advantage. This situation is different from oil transport, where tankers have the clear advantage. It reflects the high costs associated with the need to invest in LNG tankers, export and import terminals, which adversely affect the cost of transport for shorter distances.

Natural gas pipelines are used to carry gas both for domestic consumption and international trade. Natural gas for domestic use is transported from production areas to consumption areas by means of extensive national networks/grids which consist of large networks of pipelines criss-crossing many countries. The international transport of natural gas crosses land borders and seas between countries/continents, where the pipelines can link two countries directly or via transit countries.

Natural gas international pipelines are widespread, with significant networks connecting Russia with the former Soviet Union republics, with European Union countries and China; connecting the USA with Canada and Mexico; Norway with the UK and the European Union, as well as networks across South America. In the last two decades several strategic natural gas pipelines have been constructed, mainly to export the Central Asian and Caspian natural gas either westward, via Georgia to Turkey, eastward to China and southward to Iran. There are also new pipelines moving gas across Africa and from North Africa to Europe. Numerous pipelines are being planned or are under construction all over the

---

[14] EI (Statistical Review of World Energy 2024).
[15] https://blog.energybrainpool.com/tutorial-gasmarkt-teil_6-erdgastransport-und-speicherung/

world. They are mainly being established in order to connect Russia to East Asian countries, from Nigeria to Europe, from Central Asia to Europe and between Southeast Asian countries.

The practice of transporting natural gas in tankers as LNG has expanded rapidly in the last two decades. The first LNG liquefaction terminal was inaugurated 60 years ago in Algeria in 1964, with the first shipments sent to the UK and France, the expensive capital investment required to build LNG terminals has, up until recently, hindered the expansion globally. However, with increasing natural gas demand and rising prices, the natural gas market expanded rapidly. Technically, LNG trade did not exist till 1964, but it has increased steadily to achieve 59% of the global natural gas trade in 2023[16], with LNG production levels increasing by almost four-fold since 2000.[17] Impressively, in 2020, for the first time, the share of LNG gas sold surpassed pipeline trade.[18]

The main advantage of transporting natural gas as LNG using tankers, over transporting natural gas via pipelines, will be that it allows for the globalisation of the natural gas markets, rather than the current, mostly regionalised markets, where trade is only conducted via pipelines. LNG means that countries will have the choice to source their gas from far further away, if the commercial conditions are acceptable, rather than being limited to export or import natural gas based on distance constraints. Furthermore, LNG trading has brought many previously inaccessible natural gas deposits into the market.

Despite its many advantages, LNG has a major drawback, which is at this point, its relatively high cost. In its entry into the market, there are considerable initial costs involved in building the necessary infrastructure i.e. the liquefaction and regasification terminals, although the investment will be profitable within few years.

As of 2023, 21 countries had liquefaction LNG terminals, with the

---

[16] EI (Statistical Review of World Energy 2024).
[17] Ibid.
[18] https://www.gisreportsonline.com/r/gas-markets/

largest exporters being the USA, Australia, and Qatar. 53 countries have regasification terminals, with the largest importers being China, Japan, and South Korea. As the number of participating countries continues to grow each year, with more LNG terminals being built, the natural gas trade is transforming, with the possibility of expanding to become as widespread as the crude oil trade. A complete transformation is envisaged to be gradual, since a large proportion of current LNG trade contracts are still long-term and only limited volumes of LNG are available on the spot market.

1.3   Natural Gas Usage and its Role in Energy Transition

Despite all the talk, the highly emotive arguments, protests and loud debate heard everywhere around energy transition from fossil fuels to green energy, there are certain realities that need to be acknowledged. In the last four years, the world experienced significant unexpected catastrophic events. We learned that, following the Covid-19 pandemic and the conflict in Ukraine with the subsequent energy supply crunch, we need to diversify sources of fuel/energy in order to avoid supply shortages. Only by doing this, can governments establish security of supply conditions so that each country's economy can continue to function.

Thus, when the question 'Can fossil fuels can be substituted?' is asked, we need to consider the share of energy usage of fossil fuels in the overall energy demand at the present time. We need to assess if it is technically and economically feasible to replace that energy share with currently available alternatives. The numbers tell us the facts, loud and clear.

Figure 1.4 shows the world energy consumption from 1980 to 2023, where the upward trend shows no signs of reversing. 2020 was an anomaly, since due to Covid-19 pandemic, global consumption dropped.

In 2010 fossil fuels accounted for 87% of the total energy demand. Between 2010 and 2015, the share of alternative energy in the global energy mix had increased by 0.9% to 14%, whereas fossil fuels accounted for 86%, so they were still dominant. By 2019 conditions improved slightly for alternatives where their share had increased, exceeding 15%, but fossil fuels were still dominant, accounting for 85%. In 2023, the data shows that the share of fossil fuels in energy demand is declining. However, the decline is slow i.e. 2% over 13 years and, if this rate is not accelerated significantly, it will take fifty years to eliminate fossil fuels usage and dominance.

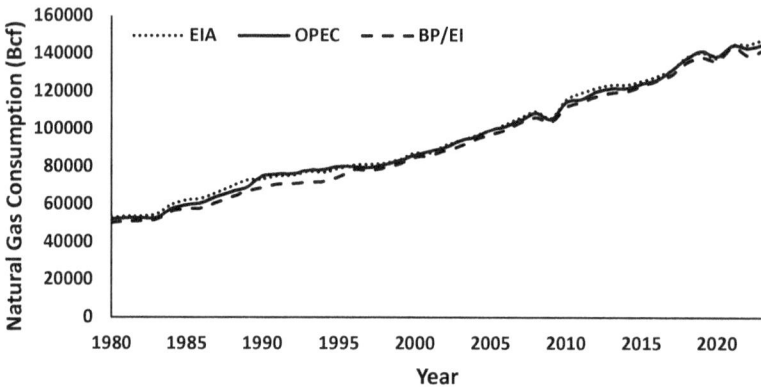

Figure 1.4: World natural gas consumption (1980 to 2023)

Source: EIA (http://www.eia.doe.gov/international); OPEC (Annual Statistical Bulletins 1999-2023); BP (Statistical Review of World Energy 2001-2022), EI (Statistical Review of World Energy 2023-2024).

Figure 1.5 shows primary energy sources consumption. As seen, natural gas has increased its share as primary energy source and is not retreating in the short term. Its main roles are in power generation, and for industrial uses. Residential and commercial, transport and feedstock/raw material are small percentages. See Figure 1.6.

The graphs below indicate the share of alternative or renewable energy increased in the energy market over time. The examination of future energy 'supply and demand' scenarios authored by governments, think tanks, research institutions, organisations and companies suggests this trend will continue. The Paris Agreement and subsequent COP summits have set out the roadmap to achieve this increase in usage, or development of alternative energy sources. At the COP28 held in the UAE in 2023, the agreed statement, although the language was weak and lacked enforcement mechanisms, called on the participating countries to reduce fossil fuel usage.

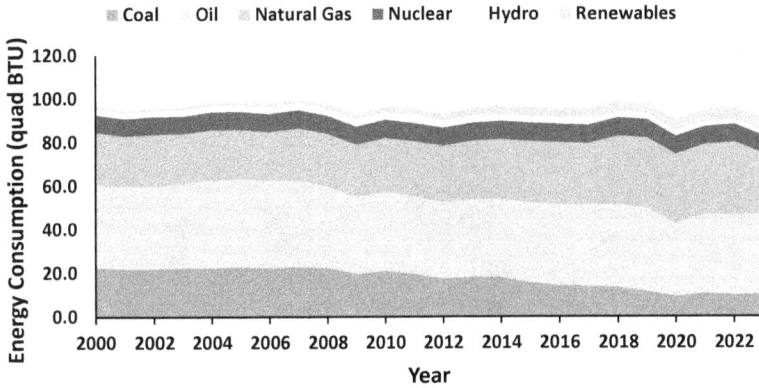

(a) Net consumption

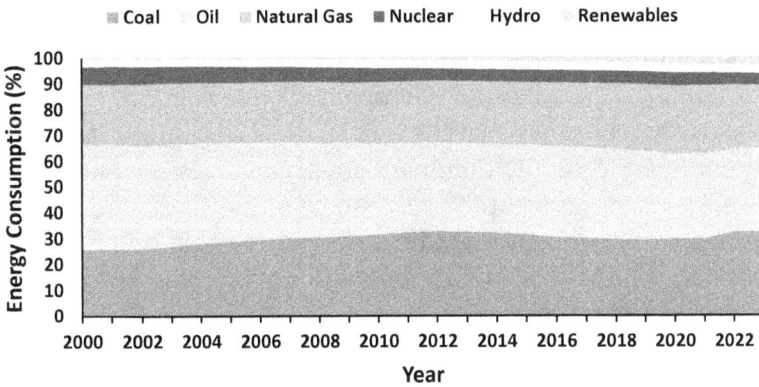

(b) Percentage consumption

Figure 1.5: Global energy mix (2000 to 2023)
Source: EIA, EI

The COP Summits roadmaps have been drawn from a variety of suggestions, opinions, and solutions offered by a host of

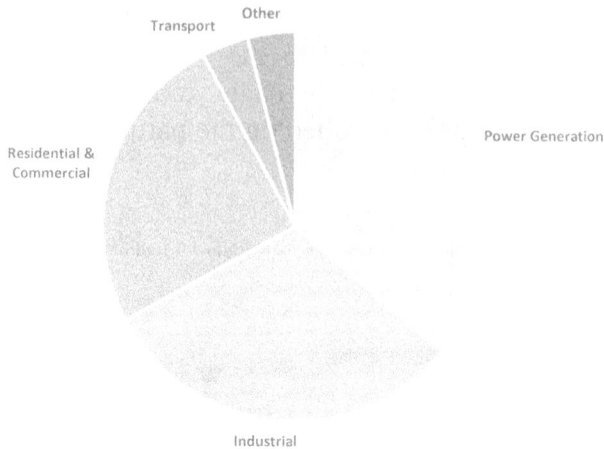

Figure 1.6: Primary Energy of gas per sector

Source: https://www.cmegroup.com/articles/2023/natural-gas-from-pipelines-to-portfolios.html

organisations, governments and companies, interestingly, the views of all these stakeholders differ significantly. I am not going to discuss their merits here, as this is outside the scope of this book. Interested readers can consult many published articles, e.g. scenarios from the IEA,[19] EIA,[20] OPEC,[21] BP,[22] TotalEnergies,[23] S&P Global,[24] Mckinsey & Company,[25] etc. There is a consensus which is that, until at least the mid of the 21st century, fossil fuels will continue to be the dominant supplying energy source. There is disagreement on what the percentage of the share will be; hence the degree of dominance will be different. Overall, renewables will account for less than 40% of the share in most future energy scenarios. Almost all energy demand and consumption scenarios, published by numerous stakeholders, agree that oil and coal usage will decline. On the other hand, the role of natural gas continues to be debated. Many scenarios suggest that it may still grow in the short and medium term, but its involvement in the energy mix will

---

[19] https://www.iea.org/reports/world-energy-model
[20] https://www.eia.gov/outlooks/aeo/
[21] https://www.opec.org/opec_web/en/publications/340.htm
[22] https://www.bp.com/en/global/corporate/energy-economics/energy-outlook.html
[23] https://totalenergies.com/media/news/press-releases/totalenergies-energy-outlook-2023
[24] https://www.spglobal.com/commodityinsights/en/ci/products/energy-climate-scenarios.html
[25] https://www.mckinsey.com/industries/oil-and-gas/our-insights/energy-2050-insights-from-the-ground-up

certainly be longer.

In a more likely scenario, natural gas may be used as a rapid solution or bridging way, to cut carbon dioxide emissions, by replacing other fossil fuels (e.g. coal) in baseload power generation, since its emissions are considerably lower. It may also support the greater rollout of renewables by providing backup capacity to overcome intermittency and reliability challenges. Natural gas can be pivotal in not only correcting imbalances of renewables' intermittent output, but also to support their rise.

Batteries can also play an important role in counter-balancing the fluctuations in output from renewable energies. In their current forms, batteries cannot provide an adequate backup alternative energy solution. While batteries can store surplus energy from renewables, the storage is only suitable to balance short-term (measured in hours) fluctuations in supply or demand of renewable energy. In order to balance days, or months, of supply and demand, a more reliable baseload solution is needed. That is where natural gas is now positioned, and combining it with carbon capture utilisation or storage initiatives, could effectively turn it into a 'green fuel', (although this label is disputed). Thus, with the natural gas market continuing to grow, it has been promoted as the bridge fuel to slash the world's oil and coal dependence, while buying time to develop alternative energy sources thereby reducing greenhouse gas emissions.

There is consensus that energy transition is happening and whoever disagrees with this is in denial. The truth is clear, as is said in this Arabic proverb 'الشمس لا تغطى بغربال', which means 'the sun cannot be covered with a sieve'. The question is not if the transition is happening, but rather with its timing. Which then leads to other important questions - when will it happen? What will its pace and its extent be? Who will win and who will lose?

In the last few years, the world has witnessed a change in the attitude of oil and natural gas companies, with many appearing to endorse a greener agenda. Companies are seeking ways to balance cost efficiencies with green measures, while maintaining their social license to operate.[26] They are setting emissions targets, have clear climate policies and net zero ambitions.[27] Many diverted some of their investments into renewable energy, moving capital expenditure (capex) from upstream, with several CEOs claiming this change will lead to the creation of new jobs and opportunities, although doubts regarding this, continue to linger.

However, there are marked differences in attitudes towards the transformation between the US and Europe,[28] as well as between international oil companies (IOCs) and national oil companies (NOCs). Despite those differences, NOCs will have to adhere to the policies of the IOCs they partner with, as they are often dependent on financial support from international financial institutions, who require certain Environmental, Social, and Governance (ESG) credentials from their creditors. NOCs, like the IOCs, can also be exposed to hostile public relations campaigns if they are not viewed as climate-friendly operators.

The 'greening' trend started with European companies, who all made acquisitions of companies in renewable energy, utilities and distribution networks, led by Equinor,[29] BP,[30] Shell,[31] TotalEnergies,[32,33] Repsol[34] and Eni,[35,36]. US companies are still way behind, remaining slow to move in this direction, with mainly Occidental and ConocoPhillips heading in this direction.[37] However, momentum accelerated recently, with other large

---

[26] https://www.hartenergy.com/exclusives/what-oil-price-crash-means-decarbonization-187423
[27] https://www.hartenergy.com/exclusives/oil-ceos-pen-open-letter-pushing-climate-change-action-187726
[28] https://www.worldoil.com//news/2020/5/12/supermajors-all-have-ambitious-and-widely-varying-net-zero-goals
[29] https://www.equinor.com/en/news/20201102-emissions.html
[30] https://www.bp.com/en/global/corporate/news-and-insights/press-releases/bernard-looney-announces-new-ambition-for-bp.html
[31] https://www.shell.com/energy-and-innovation/the-energy-future/shells-ambition-to-be-a-net-zero-emissions-energy-business.html
[32] https://www.total.com/media/news/total-adopts-new-climate-ambition-get-net-zero-2050
[33] https://www.worldoil.com/news/2020/5/5/total-pledges-to-be-carbon-neutral-by-2050
[34] https://www.repsol.com/en/press-room/press-releases/2019/repsol-will-be-a-net-zero-emissions-company-by-2050.cshtml
[35] https://www.eni.com/en-IT/low-carbon/strategy-climate-change.html
[36] https://www.edie.net/news/6/Oil-giant-Eni-targets-net-zero-carbon-emissions-by-2030/
[37] https://insideclimatenews.org/news/12112020/two-us-oil-companies-join-their-european-counterparts-making-net-zero-pledges/

players in the US and worldwide, including ExxonMobil and Saudi Aramco, joining in. [38]

Although many companies portray these actions as 'good deeds' that demonstrate their morals, ethics or commitment to social responsibilities, the reality is that, due to public and investors pressure, regulatory obligations, business necessities and realities, many companies were forced into this. The shift toward cleaner forms of energy is forcing oil and natural gas explorers to cut costs and cancel or defer plans, as uncertainties in their industry mount. In addition, digital transformation is providing a boost to energy transition, by allowing companies to reduce their carbon footprint in more efficient, cost-effective ways [39] as the new decarbonised or greener 'Energy Evolution' businesses are less capital intensive and less risky. [40]

We noticed the trend reversed in 2023, with BP and Shell scaling down their green objectives, and both were rewarded by significant increases in their share prices. [41,42]

In fact, the renewables sector was one of the few that shrugged off the devastating effects of Covid-19 and actually grew in 2020. [43] This changed in 2022 when oil and natural gas prices surged, making companies huge record profits. The renewables share of investment overtook upstream oil and natural gas in 2021 [44] and continues to expand. Although there were fears that the pandemic might slow the energy transition, [45] after an initial slowdown with

[38] https://www.theguardian.com/business/2020/jul/16/worlds-largest-oil-firm-joins-alliance-to-reduce-industrys-carbon-emissions

[39] https://eloqua.upstreamintel.com/LP=27640

[40] https://www.hydrocarbonprocessing.com/news/2020/06/eni-ceo-working-to-make-the-green-drive-irreversible

[41] https://www.investorschronicle.co.uk/news/2024/03/14/shell-slows-carbon-cuts-in-valuation-push/

[42] https://seekingalpha.com/article/4690648-bp-stock-can-run-once-they-admit-they-are-an-oil-company

[43] https://www.ft.com/content/205e5a30-4aba-4b9e-b27d-cc8a654ff684

[44] https://oilprice.com/Energy/Energy-General/2021-The-Year-Of-Renewable-Energy.html

[45] http://admin.petroleum-economist.com/articles/low-carbon-energy/energy-transition/2020/pandemic-risks-slowing-the-energy-transition

job cutting,[46] the sector rebounded, proving its resilience.[47] In fact in October 2020, there was a symbolic moment when NextEra (the world's largest solar and wind power generator) market valuation overtook ExxonMobil's market valuation.[48] However, this was all reversed in a short time, when investors realized that margins in this sector are low. Subsequently the fortunes of many of the, previously stars, of the sector declined and their valuations crashed. By end of 2023, ExxonMobil valuation was 3.4 multiples of NextEra.

1.4   Natural Gas Price Volatility

In order to understand natural gas pricing mechanism, it is essential to also have some knowledge of how crude oil is priced as the two are interlinked, although some decoupling is already occurring. It is important to point out that the markets are symbiotic, especially in terms of production. A substantial amount of natural gas is produced, as associated natural gas, when producing crude oil, equally, large amounts of natural gas liquids (NGLs) are produced (as a by-product) when natural gas is produced. This symbiosis is an important factor affecting both supply and demand markets, as well as the prices of both oil and natural gas. Oil price is often broadcasted in news bulletins and most people are familiar with parts of its terminology.

The price of crude oil refers usually to the price of a barrel of benchmark crude oil, quoted in US$ per barrel. Benchmark is a type of crude that serves as a reference price for buyers and sellers. They are used to facilitate trade since there are many variations and grades of crude oil. Three main benchmarks most often used are West Texas Intermediate (WTI), Brent, and Dubai. Many other benchmarks are also traded.

There is a differential in the price of a barrel of crude oil, based on its grade, which is determined by physical factors, such as its

---

[46] https://www.eenews.net/assets/2020/05/14/document_ew_03.pdf
[47]   http://admin.petroleum-economist.com/articles/low-carbon-energy/renewables/2020/like-a-bat-out-of-hell-part-four-renewables-set-for-rebound
[48] https://www.ft.com/content/39a70458-d4d1-4a6e-aca6-1d5670bade11

specific gravity[49] (API) and viscosity, chemical composition, such as sulphur content and geographic factors, such as its location, in particular its export routes.

Heavier crudes (higher specific gravity) and sour (high sulphur) crudes are generally priced lower than lighter or sweeter crudes, because they are harder to transport, and their refining process requires additional equipment.

Unlike crude oil, since the composition of natural gas is often homogenous, its price differs based on location, rather than physical or chemical factors. It is always priced based on its energy content and is often measured in US$ per MMBtu.[50]

Its benchmarks are often geographic hubs and are affected by the trade routes. The most important are the Henry Hub in North America, the National Balancing Point (NBP) in the UK, and the Title Transfer Facility (TTF) in the Netherlands. In addition, there are advanced initiatives to establish Asian hubs in Singapore, Japan, South Korea and China. Actual benchmark prices are determined freely, based on market forces, so significant variations between benchmarks exist and some are occasionally sold at discounts. See Figure 1.7.

As mentioned earlier, due to the ease of transportation and storage, crude oil is a global commodity that trades in markets around the world, both as spot oil and via derivatives contracts. However, unlike crude oil, natural gas is yet to become such a commodity, thus the natural gas price is not universally commodified yet. In North America, natural gas is priced as a commodity, while in Europe and Asia, the prices are often linked to the price of crude oil and/or petroleum products. Spot pricing is being gradually adopted in all regions.[51] In the last few years LNG pricing is heading into becoming commodified, but more needs to be done

---

[49] Specific gravity is the ratio of a material's density with that of water at 4 °C.
[50] One million British Thermal Units.
[51] https://www.eia.gov/todayinenergy/detail.cfm?id=23132

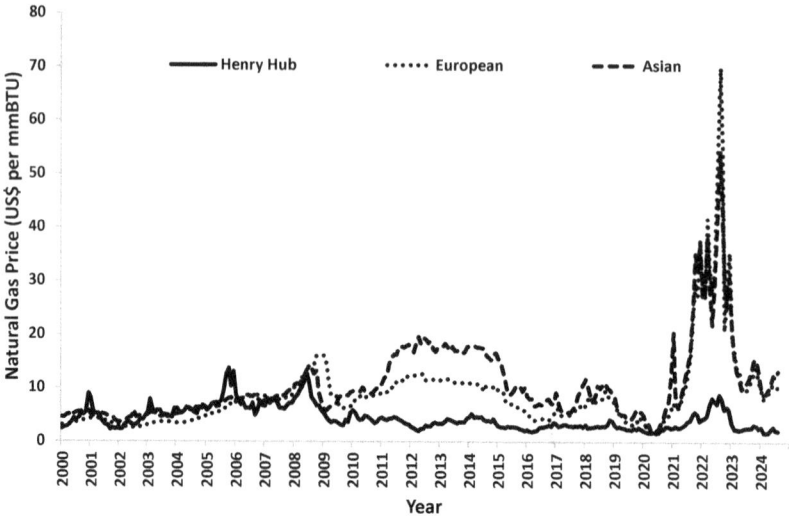

(a) Nominal price (2000 – 2024)

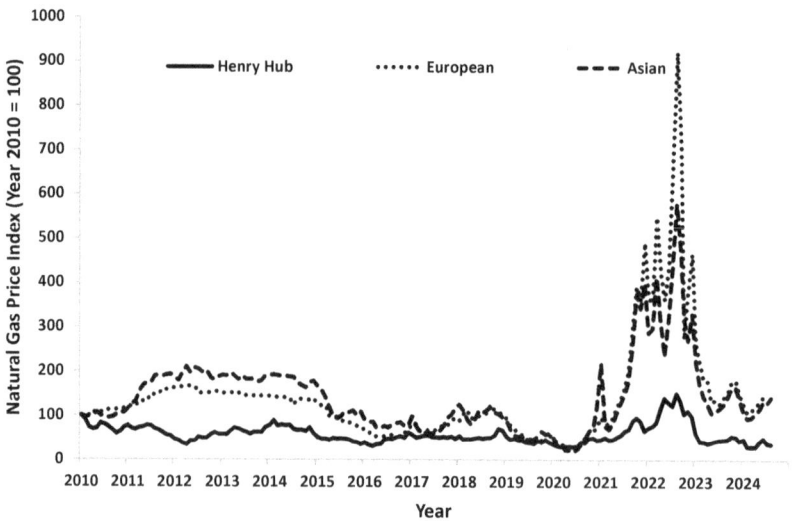

(b) Relative price to 2010

Figure 1.7: Nominal natural gas price

Source: IMF
Note 1: Prices are monthly averages.
Note 2: Prices reported till August 2024.
Note 3: Year 2010 equals 100 for relative values.

before this is fully realised.[52,53]

Historically, oil price volatility often grabbed the headlines, while natural gas price was often an afterthought. That was until mid-2021, when natural gas prices started grabbing the headlines in Asia and Europe, subsequently becoming the big story in 2022 and continuing in 2023, but fading early in 2024.

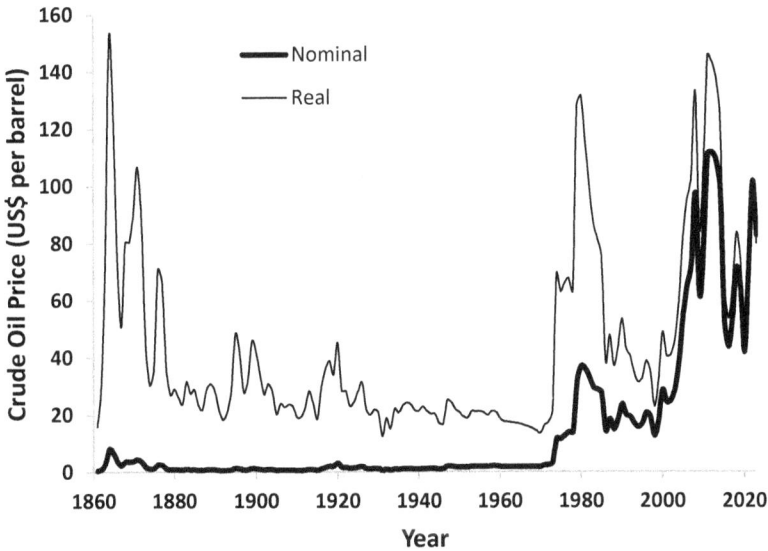

Figure 1.8: Oil price trajectory

Source: BP/EI, EIA
Note 1: Consumer price index used from
(https://www.minneapolisfed.org/community/teaching-aids/cpi-calculator-information/consumer-price-index-1800).
Note 2: Prices are yearly averages.

Starting with crude oil prices, we saw a rollercoaster ride (See Figure 1.8) that has intensified since the beginning of this century experiencing several peaks and troughs over the last two decades. See Figure 1.9.

[52] https://www.plattsinsight.com/insight/dawn-of-lng/
[53] https://www.spglobal.com/commodityinsights/en/market-insights/topics/lng-commoditization

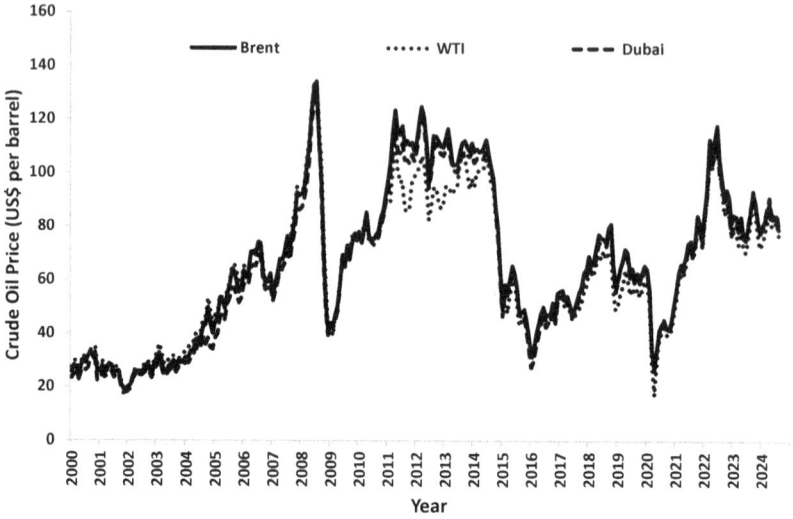

(a) Nominal price (2000 –2024)

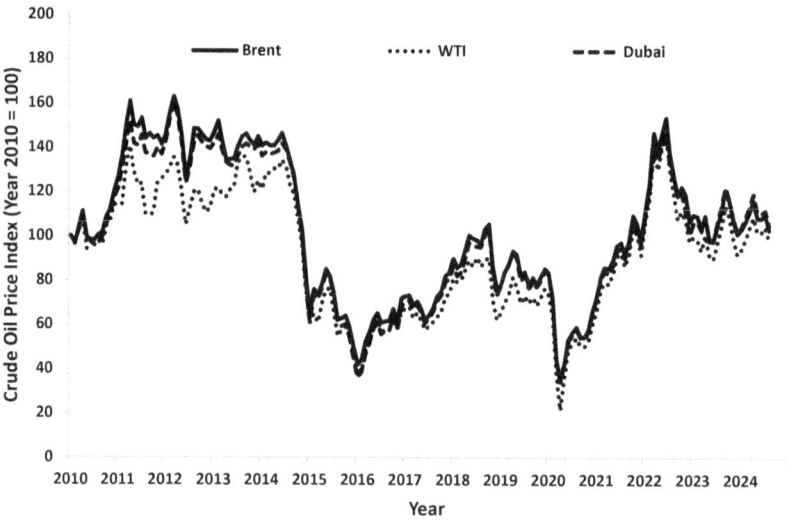

(b) Relative price to 2010

Figure 1.9: Nominal crude oil price

Source: IMF
Note 1: Prices are monthly averages.
Note 2: Prices reported till August 2024.
Note 3: Year 2010 equals 100 for relative values.

Natural gas prices also experienced a rollercoaster, but that was distinctly different. Natural gas prices continued to be regionalised, and a substantial share of the trade continues to be partially oil price indexed, with different pricing in each region. Thus, for long periods they mirrored oil prices, therefore showed volatility and collapsed in November 2014, dropping by over 50%.

During the period, 2015 to 2019, there was a flurry of natural gas projects developing several LNG export terminals, especially in Australia and the US. This additional capacity created excess supply which, coupled with low demand due to Covid-19 pandemic, exerted pressure on natural gas prices and thus bringing them down. This trend mirrored crude oil prices in 2020. The plummeting natural gas demand during the Covid-19 pandemic, threatened the return of the spectre of natural gas prices turning negative.[54,55] Subsequently natural gas prices dropped to a 25-year low,[56] which led to fears of inevitable shut-ins in production, due to the natural gas price plunge.[57,58] In early 2020 this led to record low prices and the disappearance of differentiation of regional prices (for a limited period). However, with the reduction in crude oil production, due to a significant drop in crude oil demand and subsequently price, associated natural gas volumes were cut, resulting in a tightening of the market, triggering natural gas price increases in the long term. Nonetheless, periodic spikes in price continued to occur due to certain specific events or conditions, such as cold weather or supply bottlenecks.

Thus, while the Henry Hub price never recovered to 2008 heights, Asian and European prices recovered in 2021. Cold winters in Asia in 2020-21 reversed the LNG trend price, surprising the market

---

[54] https://www.worldoil.com//news/2020/6/3/natural-gas-prices-could-go-negative-on-global-oversupply
[55] Natural gas prices reached negative previously in certain areas, such as in Texas where producers had to pay for pipeline capacity, to be able to transport their natural gas away. See https://www.ft.com/content/ff4d72e9-65b6-3466-9002-f8b561191dde
[56] https://oilprice.com/Energy/Gas-Prices/Natural-Gas-Drops-To-25-Year-Low-As-Demand-Disintegrates.html
[57] https://oilprice.com/Energy/Natural-Gas/Natural-Gas-Price-Plunge-Could-Soon-Lead-To-Shut-Ins.html
[58] https://uk.reuters.com/article/uk-column-russell-lng-asia/column-spot-lng-the-worst-performing-energy-commodity-faces-more-price-pain-idUKKBN23B0PQ

and setting record high spot prices.[59,60] Coupled with increased demand in Latin America due to lower hydropower output[61] and a reduction to natural gas supply from Russia to Europe, this subsequently led to European price increases as competition to secure natural gas supply heated up. In February 2022 the European prices spiked following the Russian invasion of Ukraine, dragging with it all other natural gas benchmarks, albeit with different levels.

In hindsight it appears that the reduction in gas volumes exported/supplied by Russia to Europe in the summer of 2021 and beyond, was a Russian plan to create an energy crisis in Europe prior to invading Ukraine, which could be used as leverage to influence Europe's reaction to the invasion and a pressure tactic to sway it from supporting Ukraine. Although the initial reaction to the invasion led to record high natural gas prices in Europe and Asia, this tactic failed spectacularly. By early 2023 we witnessed a return of prices to pre-Russian invasion levels. Europe managed to secure alternative supplies,[62] reduced its consumption of natural gas, and was blessed with a relatively warm winter.[63] Prices sank to 2 year low in June 2023, before doubling again in ten days highlighting the volatility of the price.[64] In February and March 2024, it fell to levels seen in 2021, prior to the war.[65] However, the switch from depending on pipeline natural gas from Russia, to relying on LNG mostly from the USA and the Middle East, remains risky for Europe. This vulnerability to natural gas supply means that it could still experience future price spikes and extreme volatility. Similarly, the Henry Hub price sank in 2023, in what the Financial Times dubbed as 'déjà vu all over again',[66] and the price

[59] https://shippingwatch.com/carriers/article12680458.ece
[60] https://www.spglobal.com/platts/en/market-insights/latest-news/natural-gas/011321-factbox-asian-spot-lng-prices-hit-record-highs-on-supply-glitches-demand-spike
[61] https://www.theecoexperts.co.uk/blog/reasons-for-uk-gas-price-increase
[62] https://www.ft.com/content/6f83306f-14ef-458f-a47e-1cac9c6c6dc6?desktop=true&segmentId=dd5c99e9-30be-ddd0-c634-ff3a0c2b738f#myft:noti%E2%80%A6
[63] https://www.ft.com/content/8b410d45-7dd3-4a56-a83c-8cec4bd41d6f?desktop=true&segmentId=dd5c99e9-30be-ddd0-c634-ff3a0c2b738f#myft:n%E2%80%A6
[64] https://www.ft.com/content/9265b042-6b19-460a-b946-5dd53b406544?desktop=true&segmentId=dd5c99e9-30be-ddd0-c634-ff3a0c2b738f#myft:%E2%80%A6
[65] https://www.ft.com/content/84b3ace6-c1ce-4226-a9ec-6724e56f8e68?desktop=true&segmentId=dd5c99e9-30be-ddd0-c634-ff3a0c2b738f#myft:n%E2%80%A6
[66] https://www.ft.com/content/d5d0bc54-9861-4259-9e13- d5cf616506c4?desktop=true&

plunged to a near-three-decade low in February/March 2024.[67]

Natural gas price volatility has been the market norm and continues to be so, where the regional prices exhibit ups and downs, similar to crude oil price volatility, albeit the timing of these peaks and troughs differs. Figure 1.10 shows regional natural gas prices, relative to Henry Hub since 2000, demonstrating the regionality of the pricing. During the Covid-19 pandemic, regional price differences narrowed, and in some instances, US LNG was the most expensive in the world[68] – a prospect never previously contemplated.

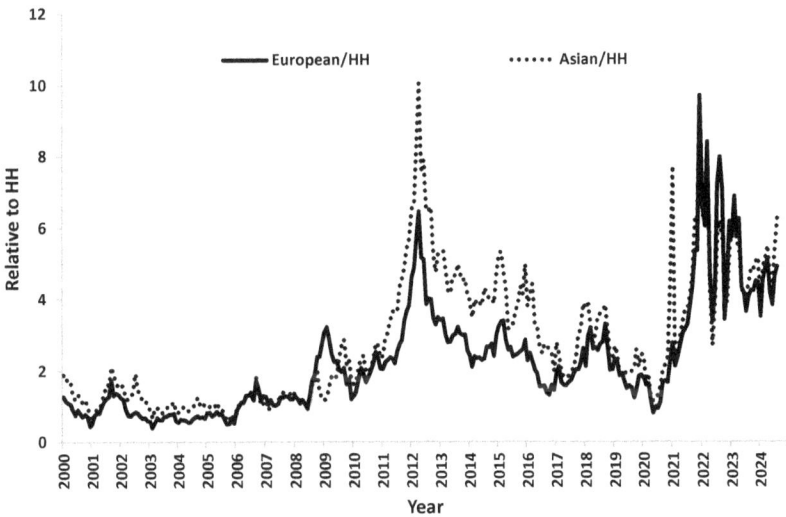

Figure 1.10: Benchmark natural gas price differentiation (2000 – 2024)

Source: IMF
Note 1: Prices are monthly averages.

Just to illustrate the disparity in natural gas prices between regions, consider the simplistic example of the valuation of the natural gas reserves of an asset of 1 Tcf. If this asset produces a tenth of its

segmentId=fe4987a4-0d36-0eb5-d88f-99ac7b30c569#myft:%E2%80%A6
[67] https://www.ft.com/content/08fa8ef9-8da9-49d0-97ab-3abbe5855d50?desktop=true&segmentId=dd5c99e9-30be-ddd0-c634-ff3a0c2b738f#myft:n%E2%80%A6
[68] https://www.hartenergy.com/news/lng-buyers-cancel-cargoes-us-natgas-becomes-most-expensive-world-187467

reserves a year, its yearly revenue would have been 278 million US$ in 2023 taking an average price of Henry Hub in North America, the same asset will generate 1.4 billion US$ if it was sold in Asian prices, and 1.34 billion US$ if it was sold in European prices. If the revenue was in crude oil prices (Brent) its value would have been 1.42 billion US$. Note that for simplicity a yearly average price of 2023 was used for the calculation.

Obviously, the above numbers are theoretical values and will not be fully realised as substantial costs and expenses are needed to develop natural gas assets including licensing, exploring, field develop, field operation and maintenance, marketing the products and paying the taxes, thus a fraction of sell price in reality is only realised.

In the last two years it is evident that, although price recovery occurred for both crude oil and natural gas, recovery trajectories after the collapse, differed significantly. There was a total disconnect between price collapses and recoveries in 2021 and prices soared, reaching record-breaking heights not seen before. See Figure 1.7 presented earlier. Thus, it is apparent that the correlation between crude oil and natural gas prices are diverging, with future price movements expected to separate.

While forecasts for future natural gas price remain low in North America, they are expected to remain elevated in Asia and Europe for the medium term. Moreover, periodic spikes in price occurred in 2023 and will occur beyond due to certain conditions, such as cold weather or supply bottlenecks. However, it is interesting to realise that the natural gas price war premium is gone, as the world has quickly learned from the wars in Ukraine and the Middle East, and managed to adapt. Although forecasting natural gas price has still a wide range of uncertainty due to these wars.

It is important to put the prices of crude oil and natural gas into perspective. Figure 1.11 shows the natural gas price as a function of crude oil price. As stated earlier, the two are now diverging, especially in North America, where in terms of equivalent heating value, natural gas is significantly cheaper. This is not always the case European and Asian customers, especially in the last two years where, in terms of equivalent heating value, natural gas price is considerably higher by several multiples.

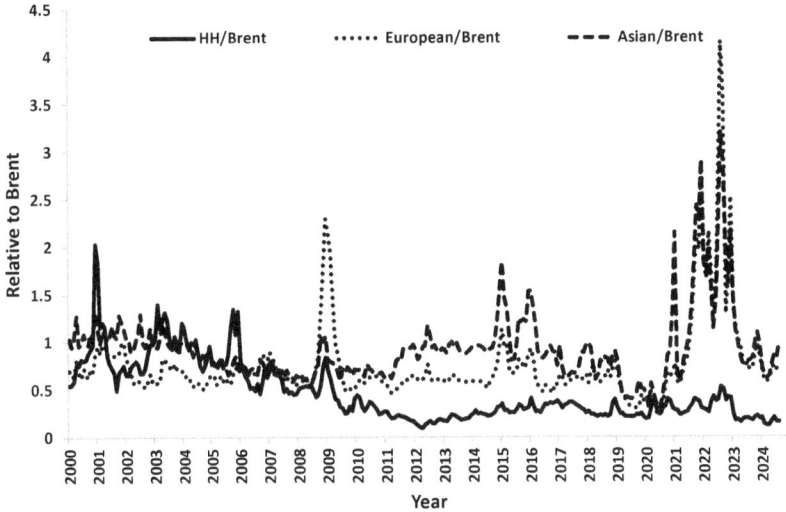

Figure 1.11: Relative nominal natural gas to crude oil price (2000 – 2024)

Source: IMF
Note 1: Prices are monthly averages.

Consequently, the link of natural gas price to crude oil price, which had already diverged, will be hard to restore due to the price volatility of both fuels, their variable demand patterns, multiple players, geopolitical situations, and economic liberalisation processes.

Chapter 2
THE PLAYING FIELD: EASTERN MEDITERRANEAN REGION

The Eastern Mediterranean region, often abbreviated as 'East Med', refers to the region surrounding the eastern part of the Mediterranean Sea, sometimes referred to as the Levantine Sea.

Figure 2.1: East Mediterranean region

Note 1: Borders are approximate and do not mean acceptance of claims

The extent of the region differs depending on how broad the definition of the region is considered. It can be very narrow, to mean only the historic Levant (Syria, Lebanon, Israel, Palestine, and Jordan), Cyprus and Turkey, which limits it to parts of Western

Asia only. Or it can be much broader, to include also adjacent, historically tied, neighbouring countries: Greece, Egypt, and the Cyrenaica province in Libya. This extends the region into three continents: Asia, Europe and Africa.

Nowadays, an even broader definition is sometimes used which includes all countries east of a line connecting Italy and Tunisia. In this extra broad definition, the whole of Libya, Tunisia, Italy, Malta and the Balkans are part of the region.

Historically, the Eastern Mediterranean countries have been military and political rivals, but they are culturally close. They share many customs and traditions, a distinct cuisine and a long, intertwined history.

Note that Jordan is often considered as part of the Eastern Mediterranean despite the fact it has no actual shores on it.

In this book, the definition of Eastern Mediterranean includes the following countries: Syria, Lebanon, Israel, Palestine, Jordan, Cyprus, Turkey, Greece, Egypt and Libya. See Figure 2.1.

A summary of vital statistics covering the ten countries can be found in Appendix I.

In the last decade, the Eastern Mediterranean region has bolstered its reputation as a major prospect in the energy supply sphere. After decades of dependency on others to supply its energy needs, the region transformed itself into a hot spot for oil and natural gas exploration. It is currently attracting lots of attention as a possible new export region.

In this chapter, I briefly discuss the political situation in the area, above ground and its geological situation below ground, in both cases hardly scratching the surface. It is clear that, what lies underground, will have an important role to play in the events occurring above ground.

2.1   Political Scene

The ten counties in the region can be characterised by rivalry and instability. I will attempt to, as simply as possible, present the status of the countries and the dynamics between them, in the bullet points below:

- Since October 2023 the region entered the unknown with the war between Israel and Hamas in the Gaza Strip. This war is proving to be the most violent between Israel and the Palestinians since the establishment of Israel. At the time of writing (June 2024) there is no prospect of the war ending.

- Israel is suffering from extreme political polarisation within, that threatens its democratic institutions and stability.

- The Arab-Israeli conflict, which is ongoing for the last 75 years, has intensified following the Israel-Hamas war, and remains an obstacle to most collaboration initiatives in the region.

- Syria and Libya are in the midst of a constant cycle of civil wars and proxy wars.

- Syria, Libya and Lebanon are officially at war with Israel.

- Turkey and Greece are effectively enemies, although they have a diplomatic relationship and are members of NATO.

- Several countries in the region do not recognise each other – Turkey does not recognise Cyprus; Syria, Lebanon and Libya do not recognise Israel; none except Turkey recognises the Turkish Republic of Northern Cyprus (TRNC).

- Cyprus is divided into an internationally recognised Greek government and the unilaterally declared Turkish Republic of Northern Cyprus (TRNC).

- Lebanon is in a complete messy situation, with collapsed economy and paralysed political system that cannot even elect a president for over a year.

- Despite the recent improvement in its economy, Greece is often at risk of economic collapse and 'Grexit' is always on the cards.

- Egypt is an authoritarian regime.

- Turkey is rapidly sliding into becoming a dictatorship.

- Although Jordan is one of the most economically vulnerable and weak countries in the region, it is relatively stable.

- Islamic insurgency and extremism are spreading, threatening further instability in the region, which has implications reaching beyond the region's borders.

- Greece has a regional maritime alliance with the Greek Cypriot administration, Egypt and Israel. [1] There is another separate alliance between Turkey, the Turkish Cypriot administration and the Libyan government in Tripoli.

- Most countries in the region have cultural rivalries. They disagree on many things, like food, e.g. who invented falafel, hummus and shawarma (all traditional foods in the region).

- While energy appears to be the root of most tensions, conflicts and infights in the region, water will be the resource that ignites the region. This region suffers from

---

[1] https://www.aa.com.tr/en/politics/turkish-libyan-maritime-pact-a-game-changer-in-emed/1671447

severe water scarcity, which is being amplified by the effects of climate change, population growth and over-extraction of water resources. The region is one of the most water-stressed in the world, with countries like Jordan, Israel facing severe water shortages. In fact, water was one of the major contributors to previous conflicts in the region, as on 1967 war.

- Suez Canal is very strategic to world trade and supply chains, and thus its security is vital. The internet is rife of baseless conspiracy theories that Israel plans to build a competitive Red Sea – Mediterranean Sea Canal to undermines Suez Canal.

- Israel is a nuclear power, raising the possibility of a nuclear arms race in the region.

- Turkey's location made it a crossroad of pipelines, and Russia wants to turn it into a natural gas hub, which is likely to cause dissent with Europe.

- Egypt wants to capitalize on its current natural gas infrastructure and the Suez Canal and wants to transform itself into a natural gas hub.

Quantitatively three international indicators summarise the status of the region's countries. These are the fragile states index (FSI), the human development index (HDI) and the freedom world index (FWI).[2]

Starting the analysis, we examine the first indicator, the fragile states index (FSI). This came into existence in 2005, when the Fund for Peace[3] produced the first Failed States Index (FSI)[4]. The index

---

[2] Note that all the numbers for Palestine in the following discussion are prior to the total destruction of Gaza Strip by Israel in 2023-24 war.
[3] The Fund for Peace is a non-governmental research and educational institution, which researches ways to prevent violent conflict and promote sustainable security. It was founded in 1957, and is based in Washington DC. (www.fundforpeace.org).
[4] https://fragilestatesindex.org

was later renamed Fragile States Index (FSI) in 2014. This was attributed to scholarly criticism of the term 'failed', where some claim it was incorrect terminology. Because the term is in the past tense, critics claimed it suggests and implies that failure is everlasting and there is no way back.

Since its inception, FSI is published jointly by the Fund for Peace with the Foreign Policy Magazine, until 2018 and then with the New Humanitarian[5]. The index covers 179 countries, scoring each country, using on twelve social, economic, political and military indicators. It spans an overall score between 0 and 120. It then categorises States into several categories, which evolved over time and are currently ranging from 'very high alert' in the bottom category, to 'very high sustainable', in the top category.

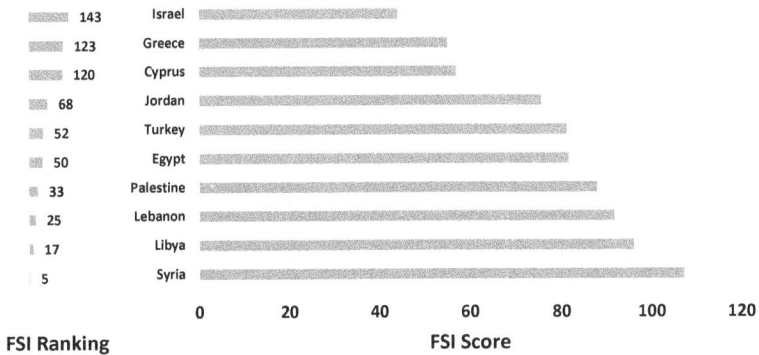

Figure 2.2: Fragile States Index (FSI) in the Eastern Mediterranean countries 2023

Source: The Fund for Peace, 2024 Edition
Note 1: Scores are interpreted as <30 Sustainable, 30 - 69.9 Stable, 70 – 89.9 Warning, 90 – 120 Alert.

Figure 2.2 summarises the situation in the ten Eastern Mediterranean countries, three countries fall into the high alert category, the worst in the index, none are in the sustainable categories (i.e. the highest categories) and only three countries are in stable categories, the second best. Putting this in context, the USA's FSI score is 45.3 with a ranking of 141st and classification

---

[5] https://www.thenewhumanitarian.org/

as stable, while the UK's FSI score is 41.9 with a ranking of 148th and its classification is also stable.

Thus, with all the problems listed above, the overall stability of the region's countries are at risk, and many of these countries are considered fragile states.

Next, the second indicator, which examines the Human Development Index (HDI) in order to numerically measure human development in the region. The HDI is a statistical composite index that measures human development, by combining indicators of life expectancy, educational attainment and income to measure the impact of economic policies on people's quality of life. It is used to distinguish whether a country is developed, developing or underdeveloped.

The HDI was first published in 1990 by the United Nations Development Programme (UNDP). The UNDP identifies three foundations for human development: to live a long, healthy and creative life; to be knowledgeable (in terms of access to a school education), and to have access to resources needed for a decent standard of living.[6] These represent three main dimensions of human life: health, education and economic prosperity. Their indicators were used to calculate earlier versions of the HDI. Over the years, however, in response to certain criticisms, the calculation formula has been enhanced by making some modifications and refinements to the index.

The latest edition of the Human Development Index Report was published in 2024. It covers 193 countries and territories. Each country gets a calculated score between 0 and 1 based on defined calculation methodology.[7] Countries fall into four broad human development categories: Very High Human Development, High Human Development, Medium Human Development and Low Human Development. Based on their scores, countries are ranked;

---

[6] http://hdr.undp.org/en/humandev
[7] Note that because of the new methodology adopted since the 2010 Human Development Report, the new reported HDI figures from 2010 appear lower than the HDI figures in previous reports published prior to the new methodology.

the country's rank indicates its position, relative to the other countries and territories included in the index. Full details on the methodology and indicators used can be found on UNDP website.[8] Several countries and territories not covered by the UNDP report, e.g. Taiwan, who calculate their own HDI score using the UNDP's methodology.

Figure 2.3 shows the HDI for the ten countries in the Eastern Mediterranean, demonstrate that the region is highly developed. The latest set of results classifies four countries (Israel, Cyprus, Greece and Turkey) as very high developed, with the rest as high developed, except Syria, which witnessed a significant drop to medium, following over a decade of civil war. Still its ranking exceeds 36 countries and territories.

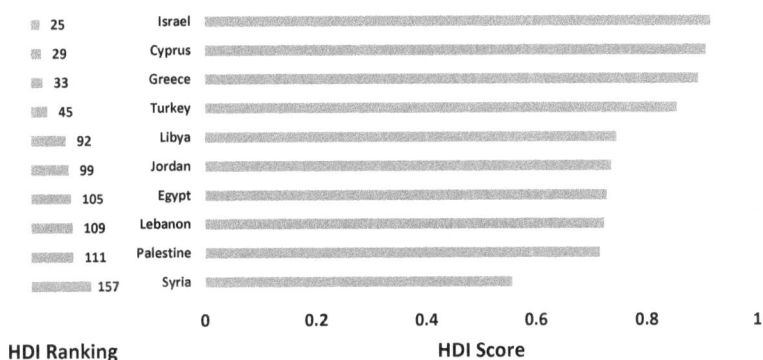

Figure 2.3: Human development index (HDI) in the Eastern Mediterranean countries 2022

Source: UNDP, 2024 Edition

Let's put the above numbers into context. The USA's HDI score is 0.927 with a ranking of 20th and classification as 'very highly developed', while the UK's HDI score is 0.94 with a ranking of 17th and classification as very highly developed.

Despite all the instability issues, when examining the human

---

[8] http://hdr.undp.org/en/humandev

development situation, all the countries in the region fare well, even Syria, despite its score dropping massively since the civil war, still outranks many in the world, mostly in Africa.

The third indicator examined is the Freedom World Index (FWI), which is one of the indicators measuring political or economic freedoms. In a nutshell, political freedoms are restricted in the region. While economic freedoms are relatively advanced, freedoms in the media are not. Internet censorship is rife, and, in most countries, opposition is discredited, often labelled as unpatriotic or criminal.

As said earlier, to quantify a measure of the political systems, the FWI is used. The FWI was devised by Freedom House,[9] based in Washington DC and has been published annually since 1972. The choice was made to include this as an indicator because the FWI has the longest chronological series and includes the largest number of countries (210 countries and territories).

The FWI researches several categories and indicators, then based on detailed analysis, it produces annual scores representing the levels of political rights and civil liberties of individuals in each state and territory, on a scale from 'most free' to 'least free'. Then the pair of political rights and civil liberties ratings is averaged to determine an overall status of 'Free', 'Partly Free' or 'Not Free'.

As can be seen from Figure 2.4, three countries are classified as 'free' – EU members Cyprus and Greece, and Israel. The Turkish Republic of Northern Cyprus is also classified as 'free'. Lebanon, despite its fragile political and economic situation, is classified as 'partially free', while all other countries are classified as 'not free'. Syria is ranked 207th out of 210 countries and territories.

However, despite the above score, the real picture is different. Only two of the 'free' countries (Greece and Cyprus) are true democracies and Cyprus is divided into two entities. The third

---

[9] Freedom House (FH) is U.S. Government funded non-governmental think tank researching democracy, political freedom, and human rights. It was formed in 1941 and is based in Washington DC. (www. freedomhouse.org).

'democracy' i.e. Israel, effectively has an apartheid system.[10] Amongst the other 'partially free' countries, one (i.e. Turkey) is a flawed pseudo-democracy, while the others are democratic in name only, in reality they are actually authoritarian states. Although the degree of authoritarianism varies significantly, where some have some window of freedom, in certain aspects and thus, the term is relative here.

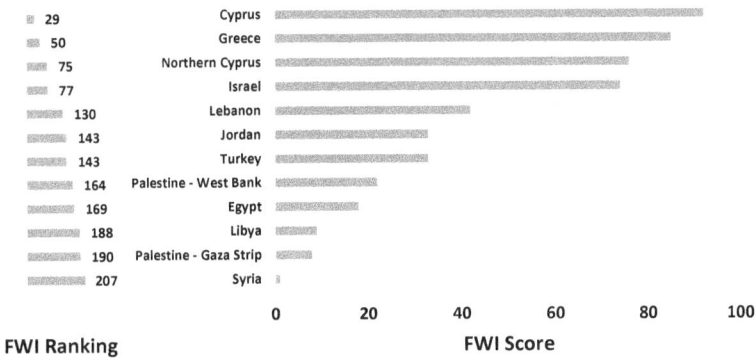

Figure 2.4: Freedom world index (FWI) in the Eastern Mediterranean countries 2023

Source: Freedom House, 2024 Edition
Note 1: Scores are interpreted as >70 Free, 35-70 Partially Free, < 35 Not Free.

To put the above numbers in context, the USA's FWI score is 83 with a ranking of 57th and classification, for the moment, is 'free', while the UK's FWI score is 91, with a ranking of 34th and classification as 'free'.

2.2   Hydrocarbon Systems in the Eastern Mediterranean

Oil and natural gas are found in systems referred to as basins, each with multiple reservoirs made of sedimentary rocks, which often span across political boundaries. They exist in geological formations, often with certain geological and physical characteristics.

The Eastern Mediterranean region contains several basins that

---

[10] The situation is worsening as a result of the ongoing war in Gaza.

overlap political boundaries, with each basin containing both hydrocarbon prospects and several fields. These terms are used in oil and natural gas exploration terminology. The more detailed geological definitions are outside the scope of this book, however for the sake of clarification and completeness, these terms are explained in basic English in the following lengthy paragraph.[11,12]

A hydrocarbon ***province*** is a geographic term defined as an area where hydrocarbon occurs in commercial quantities. A province encompasses several geological structures including basins. A ***basin*** is a large, natural depression on the Earth's surface in which sediments accumulate, it is a complex system of hydrocarbon source rocks, migration conduits, carrier beds and traps, that contain hydrocarbon reservoirs. The terms basin and province are sometimes, confusingly, used interchangeably in literature, where provinces are referred to as basins and other geological structures, such as fold-belts which could be structural provinces, are referred to as basins. Basins can be located on top of each other at different depths. Each basin often contains multiple ***play***s, which are areas of hydrocarbon accumulations or prospects, with similar characteristics. Each play contains several ***prospect***s. A prospect is a certain location in a play with a consistent set of geological criteria and conditions, which can be feasible, economically, to justify capital investment for an exploratory well. A ***field*** is a geographical area in which one or more wells produce oil or natural gas. Each field includes one or several ***reservoir***s, which are porous and permeable underground formations trapped by barriers of impermeable rocks or water that contain accumulations of producible hydrocarbons which can be separated, either horizontally or vertically.

Thus, in terms of basins and other Hydrocarbon bearing geological structures in the Eastern Mediterranean, the number of them varies, depending on the extent of the geographical definition of the region and whether only the offshore area is considered or if both the onshore and offshore areas are included.

---

[11] https://ektinteractive.com/exploration/oil-gas-exploration-terminology/
[12] https://wiki.aapg.org/Province,_basin,_system,_play,_and_prospect

In a report by the EIA [13] that narrows the region definition to the Levant and Cyprus, assessing oil and natural gas in the Eastern Mediterranean region, eight significant 'basins' are listed: Cyprus Basin, Eratosthenes High, Latakia Basin, Levantine Basin, Judea Basin, Nile Delta Basin, Western Arabian Province and the Zagros Province. These 'basins' vary considerably in their area, resource size and composition. While the majority of historical hydrocarbon production in the region is from the Nile Delta Basin, the Western Arabian Province and the Zagros Province, most of the area of Zagros Province is located outside the East Mediterranean countries, instead being found within Gulf countries.

At the moment, there is mostly focus is on the basins located offshore, where the majority of current exploration activities are happening. Although these include Cyprus Basin, Eratosthenes High, Latakia Basin, the main focus is on the Levantine and Nile Delta Basins. The Nile Delta Basin is located mostly within Egypt's territorial waters, it is where much of Egypt's offshore oil and natural gas is produced. The Levantine Basin, where most of the focus is, comprises all the territorial waters of Israel, the Palestinian Authority, Lebanon, Syria and part of Cyprus.

Major discoveries were made in the Eratosthenes High Basin in the last few years, although most press reporting have not made a distinction between discoveries in this geological structure and the two neighbouring major basins of the Nile Delta and Levantine.

Note that the Western Arabian Province covers large parts of Jordan and Syria, extending into Iraq, Saudi Arabia and Turkey. Most of Jordan's fields and many of Syria's fields are in the Western Arabian Province.

Several other of Syria's largest fields are part of the Zagros Province, but the vast majority of fields in the Zagros Province are in other countries, including Iraq, Iran and Saudi Arabia. While the

---

[13] https://www.eia.gov/international/content/analysis/regions_of_interest/Eastern_Mediterranean /eastern-mediterranean.pdf

Western Arabian and Zagros Provinces account for most of the historical hydrocarbon production in the region, these basins are now mature. Moreover, the Zagros Province extends from Turkey in the north, through Iraq and Iran, terminating in the Gulf of Oman in the south. Most of the focus today is on the Levantine Basin, which has the most prospectives in the region.

So far, activities in other basins have rendered very little results, thus, there has been only lukewarm interest in investing in these basins by oil and natural gas companies.

2.3 Gas Discoveries in the Eastern Mediterranean

In the last two decades, the discovery of natural gas in the Eastern Mediterranean region has transformed energy politics there. While earlier discoveries transformed natural gas markets in Egypt and Israel, additional discoveries in the last few years have bolstered the Eastern Mediterranean region's reputation as a major new hydrocarbon province, transforming it into a promising area for oil and natural gas exploration. In the following paragraphs I summarise the developments in the last few years.

Until the end of the 20th century, Libya and Egypt were the only natural gas exporters in the region, albeit Libya's were from fields located, geographically, in the Western Mediterranean (see Section 3.10 for more details). Both countries supplied their neighbours via a network of pipelines and sold LNG on the international market. However, Libya's production collapsed due to its civil war and numerous conflicts, while Egypt overstretched its resources and mismanaged its fields becoming a natural gas importer in 2015 (see Section 3.1 for more details). Fortunately for Egypt, a massive natural gas discovery in Eastern Mediterranean, the Zohr field, offshore Egypt allowed it to briefly become an exporter again.

Beyond Egypt, between 1999 and 2000, several small natural gas discoveries were made in Israeli and Palestinian territorial waters. These were followed by significant discoveries in Israeli waters allowing it to become self-sufficient and turning the country into natural gas exporter.[14,15,16] (see Section 3.2 for more details).

The discoveries of natural gas reserves in Israel have certainly changed the political and economic atmosphere in the region, indeed, for a while they appeared to have altered the energy strategy in the region, where Israel had turned into a producer, and subsequently suppling natural gas to both Jordan and Egypt. It is remarkable how the situation altered in a very short time. Israel, after being vulnerable in terms of securing its energy supplies, and relying on Egypt for its natural gas needs, has become an indispensable source of natural gas to Egypt, which is now struggling to secure its natural gas supplies. This reversal of roles makes Egypt vulnerable both politically and economically. Certainly, the flow of investment in upstream energy has changed significantly for all the countries involved.

Elsewhere in the Eastern Mediterranean, Cyprus has made significant discoveries of natural gas offshore (see Section 3.4 for more details) and offshore drilling activities are also underway in Lebanese waters (see Section 3.6 for more details). Natural gas exploration in Syria was interrupted by war in Syria, and although agreements with Russian companies looked like restarting these explorations, the Russian invasion of Ukraine and subsequent international sanctions on Russian companies, delayed them again.

The discovery of Cypriot natural gas has become an issue of conflict between Turkey, Greece and the Greek Cypriot government. The Turkish Republic of Northern Cyprus claims the rights to these natural gas resources, while mainland Turkey is pursuing its own exploration activities in the region. There has been no success so far in the Eastern Mediterranean, but Turkey has had major success in the black Sea.[17] (see Section 3.5 for more

---

[14] https://www.reuters.com/article/us-israel-natgas-leviathan-idUSKBN1YZ0H9
[15] https://www.spglobal.com/platts/en/market-insights/videos/market-movers-europe/011821-nord-stream-oil-energy-forum-iea-sustainability-week-french-nuclear
[16] https://www.offshore-mag.com/production/article/14074908/leviathan-gas-flowing-to-jordan
[17] Despite recording no success in the Eastern Mediterranean waters Turkey announced a major discovery exceeding 14 Tcf of natural gas in the Black Sea (https://oilprice.com/Latest-Energy-News/World-News/Turkey-Expands-Gas-Drilling-In-The-Black-Sea.html; https://www.upstreamonline.com/exploration/turkey-increases-tuna-1-gas-catch-to-14-3-trillion-cubic-feet/2-1-895387)

details). The Cypriot dispute is complex, as several countries' territorial claims overlap. In addition, Turkey's involvement in Libya has further complicated matters with yet more contradictory territorial claims. This has created tensions between Turkey and the EU, as there is significant French involvement in the affairs of the region.

Egypt, Israel, Lebanon, Syria, Turkey, Libya and possibly Greece, have all potential resources that may yet be discovered in their territorial waters. While the discoveries to date are significant, so far, they fall short of the 122 Tcf in the Levantine Basin[18] and 223 Tcf in Nile Delta Basin[19], estimated to exist by the US Geological Survey in 2010.

Egypt is currently pursuing an ambitious plan to become a regional natural gas hub, exporting Israeli and possibly future Cypriot natural gas. It is also involved in an expensive (possibly unnecessary)[20] plan to construct an offshore natural gas pipeline for the transport of Israeli and Cypriot natural gas resources to Europe.[21] Although the pipeline plans are progressing on paper, there are complex obstacles, and the plans may prove hard to bring to fruition.[22]

Turkey is firmly opposed to these efforts. It prefers export routes that cross Turkey, which are more feasible and also promotes Turkey as the rival regional natural gas hub. However, with high tensions between Turkey, its neighbours and the EU, the situation remains uncertain. However, relations between Egypt and Turkey have improved, so who knows? Despite seventy five years of constant war, peace, progress, more conflict and destruction, which does not look like it will change anytime soon. Still where we see political alliances and international upheaval so often shift – maybe anything is possible.

---

[18] The Middle East conflict over natural gas, Ahmed Al- Bassosy, Regional Center for Strategic Studies' monthly publication of the "State of the Region"
[19] https://pubs.usgs.gov/fs/2010/3027/pdf/FS10-3027.pdf
[20] https://euobserver.com/green-deal/149929
[21] https://www.nsenergybusiness.com/projects/eastern-mediterranean-pipeline-project/
[22] https://www.dw.com/en/eastmed-gas-pipeline-flowing-full-of-troubling-questions/a-51871424

In Chapter 3 that follows we discuss in detail the natural gas status in the region, visiting each country separately.

Chapter 3
THE MAIN PLAYERS AND THEIR STRATEGIES

Such are the potential discoveries in the last decade many international oil and natural gas companies have committed significant capital expenditure (capex) to explore and develop hydrocarbon assets in the Eastern Mediterranean region.

In order to clarify the situation in the region and provide insights for the potential future outcomes of how natural gas assets will develop there, let us explore each of the ten main countries, referred to as players, in the region. For each player I will examine this from two different perspectives:

1. The opportunity of each of these countries to become a major exporter of natural gas. Following the European policy of excluding Russian natural gas imports after the invasion of Ukraine and the decoupling of the European-Russian energy contacts, this issue has gained additional importance for natural gas markets.

2. The impact of each country's increasing domestic energy consumption on where to source its natural gas as an importer and/or its effect on its ability to export natural gas (especially by Egypt, Saudi Arabia, Iraq, Algeria).

The analysis that follows takes into account complex issues where the interplay between energy markets and geopolitical situation can affect any forecast. Remember that forecasting is an art and not a science. While many planners pay attention to as many forecasts and scenarios of the future as possible, forecasters must, by necessity, revise them and assumptions used in formulating forecasting models often change. Just compare forecasts from IEA or OPEC in last 5 years, see how different they are. (See Chapter 1).

Despite the interest in this region, it is essential to realize that in energy terms, the Eastern Mediterranean Region is a minor one. Its oil and natural gas reserves are relatively small, compared to the neighbouring Gulf region. Its strategic importance stems from its location as a major transport hub.

However, despite the natural gas reserves and production's relatively small volumes, the current local market in the region will not be enough to consume all its potential natural gas production. There is a need to find new markets, which can be done by finding new markets or countries as customers and by replacing or displacing other energy sources already being consumed, such as oil or coal.

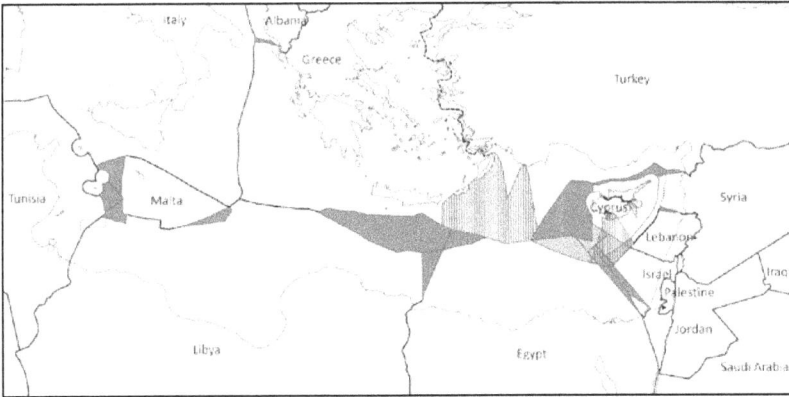

Figure 3.1: Eastern and Central Mediterranean EEZ borders, highlighting disputed areas

Source: Drawn by author based on different sources.
Note 1: Borders are approximate and do not mean acceptance of claims.

As mentioned earlier, geopolitical constraints influence natural gas policies and, before deep diving into each country, just take a look at Figure 3.1 which shows the disputed areas in the region. They say 'a picture is worth a thousand words' and this graph clearly illustrates the potential conflicts and ignition points that could affect natural gas development in the region over the next few years.

3.1   Egypt

Egypt is the largest of the Eastern Mediterranean countries, in terms of population and second only to Libya, in terms of area. This debatably largest Arab country. is strategically located at the crossroads of three continents and controls the Suez Canal, a major maritime route, vital for the flow of global trade. Historically, the various incarnations of Egypt at times, controlled other countries also in the Eastern Mediterranean, including the Levant, parts of Libya and parts of Greece. The current Egyptian territory is a mere 40% of its 1914 Sultanate territory, after losing the Sudan (modern day Sudan and South Sudan) in 1956. In the last few years, the Egyptian economy has deteriorated significantly and the Egyptian authorities' efforts to arrest the downward spiral has been unsuccessful so far.

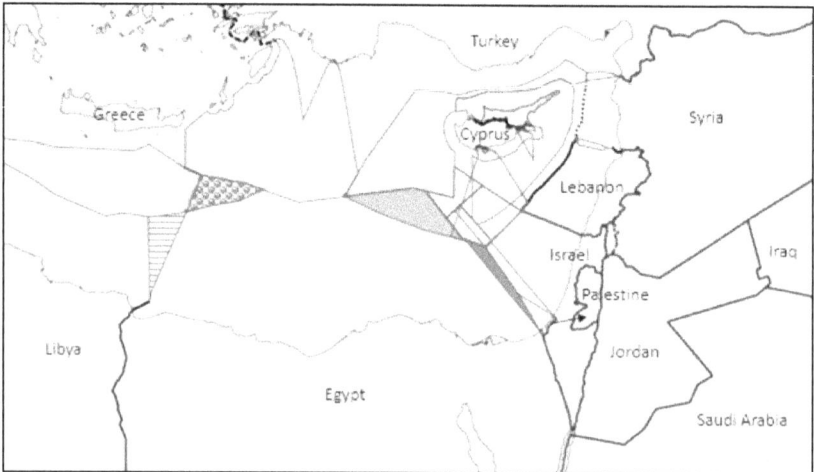

Figure 3.2: Egypt's Mediterranean EEZ, highlighting disputed areas

Source: Drawn by author based on different sources.
Note 1: Borders are approximate and do not mean acceptance of claims.

Egypt has long coasts on both the Mediterranean Sea and the Red Sea. Its exclusive economic zone (EEZ)[1] in the Eastern Mediterranean is currently governed by bilateral agreements or unilateral declarations. However, these claims are affected by other conflicting declarations and agreements from neighbouring countries. Figure 3.2 shows the current situation, summarised as follows:

1. Egypt-Cyprus EEZ is mostly covered by the 2003 bilateral agreement between Egypt and the Greek Cypriot government. The tripoint and western section potentially bordering Greece are not yet determined as the border between Greece and Cyprus is not finalised yet. That area also has conflicting claims involving Turkey. Similarly, the tripoint and eastern section potentially bordering Israel and Palestine, is not yet determined as the border between Israel and Palestine is not finalised yet. In addition, that area has conflicting claims between Egypt and Palestine.

2. Part of the maritime border between Egypt and the Palestinian Authority is governed by the interim 1994 Oslo agreement (between Israel and the Palestinian Liberation Organisation (PLO)), that established the Palestinian Authority. North of this border, there is an overlap of claims between Egypt and Palestine. At the time of writing, the war between Israel and Hamas-led Gaza is raging, and the future control of Gaza Strip is unclear, and hence the future borders between Egypt and the Palestinian Authority remain fluid.

3. Depending on the outcome of the Egyptian-Palestinian EEZ overlap claim, there is a potential maritime border between Israel and Egypt. In reality, since Israel controls

---

[1] An exclusive economic zone (EEZ), as prescribed by the 1982 United Nations Convention on the Law of the Sea, is an area of the sea in which a sovereign state has exclusive rights regarding the exploration and use of marine resources, including energy production from water and wind. EEZ does not define the ownership of any maritime features (islands, rocks and low-tide elevations) within the EEZ.

Palestine's maritime borders, this is the actual de-facto situation; however, a final formal settlement is yet to be agreed. Considering the status of the Palestinian-Israeli conflict and the raging war in Gaza in 2023-present this cannot be decided for the foreseeable future.

4. Part of the Egyptian-Greek maritime borders and EEZ delimitation is governed by a bilateral agreement of 2020. The rest is governed by the equidistant or median line principles, but it is disputed by Libya and Turkey. Note that, if Turkey and Libya succeed in their EEZ claims, there will be no maritime borders between Greece and Egypt.

5. If Egypt accepts Turkish unilateral EEZ claims, Egypt will have shared maritime borders with Turkey and will gain considerable areas from both Greece and Cyprus.[2,3] Some in Egypt advocate accepting the Turkish claim as it is beneficial to Egypt. However, while this is not the official stance yet, the Egyptian government may U-turn on its current position.

6. The Egypt-Libya maritime border and EEZ is governed by Egypt's unilateral declaration of 2022.[4] This is disputed by both Libya (who considers the borders governed by equidistant or median line principles) and by Turkey, who advocates 'equitable' principles, disregarding the rights of islands to claim EEZ. Consequently, this dispute means there is an overlap EEZ claim between Egypt and Libya.

Egypt's EEZ claims in the Red Sea are more straightforward and the maritime borders are delimited with Israel, Jordan and Saudi Arabia, following the transfer of the islands of Tiran and Sanafir to the latter. However, the maritime borders are still in dispute with Sudan, due to the disputed area of Halaib Triangle, which is

---

[2] https://www.aa.com.tr/en/energy/energy-projects/egypts-eez-could-extend-via-maritime-deal-with-turkey/32096
[3] https://stratejikguvenlik.wordpress.com/2022/11/29/dis-politika-dosyasi-cihat-yayci-yunanistani-cildirtan-libya-antlasmalari/
[4] https://greekcitytimes.com/2022/12/14/egypt-illegal-turkish-libyan/

currently under de-facto Egyptian control. This political topic is outside the scope of this book.

As the Eastern Mediterranean country with the largest population, Egypt has always been considered as the regional heavy weight. As a result of this, the current government is pursuing policies to cement its position, endeavouring to make it the cornerstone of the region's energy infrastructure. It has declared its intention to be the region's natural gas hub, although this places it in direct competition with Turkey, who is arguably more successfully attempting to achieve the same goal. (See Section 3.5).

Although the country has significant crude oil and natural gas reserves, Egypt has diversified its primary energy usage significantly in the last two decades. It has moved steadily to rely on natural gas to generate electricity, as well as a primary source of domestic and industrial fuel. After years of growth, the country's primary energy demand has steadied in the last decade. Figure 3.3 shows the energy mix and the trends, illustrating the growing role of natural gas in the energy mix, at the expense of oil.

Egypt has a long history in oil and natural gas exploration. This began as early as 1883, indeed at the time it was exploring for oil, because natural gas was considered an undesirable, that was often discarded, flared or vented. The first oil discovery was made in 1886 at Ras Gemsa, in the Eastern Desert, on the west coast of the Red Sea.[5] This was followed by many onshore oil discoveries. In 1961 Belayim, offshore in Gulf of Suez, was discovered as the first offshore oil field in Egypt,[6] since then, hundreds of discoveries were made, both on and offshore.

At the end of 2023, Egypt's proven oil reserves were 3.3 billion barrels,[7] the fifth largest in the Continent of Africa and ranked 30th

---

[5] https://egyptoil-gas.com/features/more-than-a-100-year-journey/
[6] https://www.petroleum.gov.eg/en/gas-and-petrol/discovery-search-production/Pages/Petroleum.aspx
[7] https://www.ogj.com/general-interest/economics-markets/article/14302481/global-oil-and-natural-gas-reserves-both-increase

globally.[8] Its production in 2023 was 660 thousand bbl/day.[9] Figure 3.4 shows Egypt's annual oil production and consumption, illustrating that, since 2006, the country turned into a net oil importer. Although Egypt joined the Organization of Arab Petroleum Exporting Countries (OAPEC) in 1973, it has never joined the Organization of Petroleum Exporting Countries (OPEC).

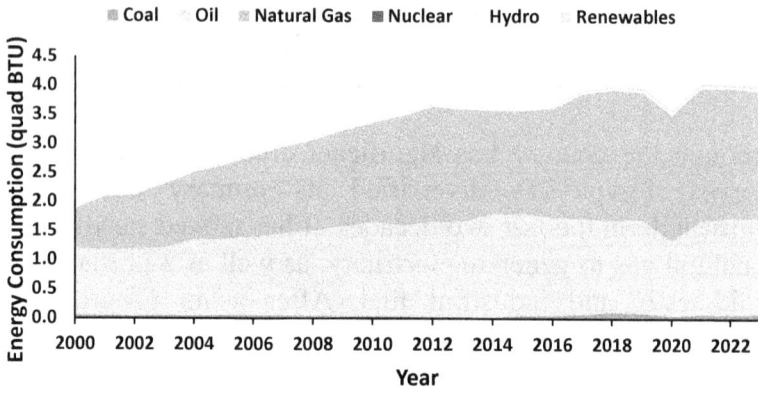

(a) Net consumption

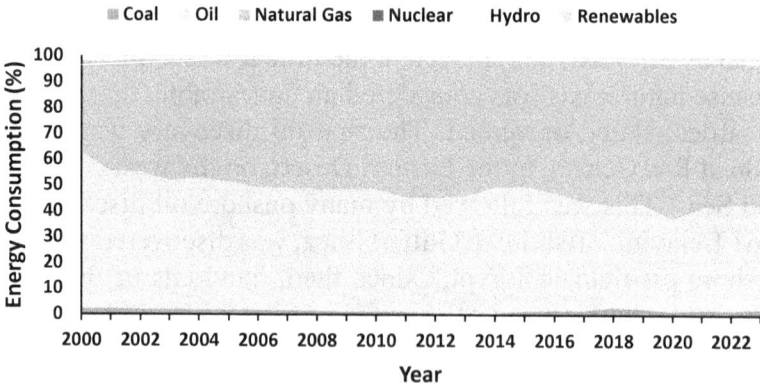

(b) Percentage consumption

Figure 3.3: Egypt energy mix (2000 to 2023)

Source: EIA, EI

---

[8] https://en.wikipedia.org/wiki/List_of_countries_by_proven_oil_reserves
[9] https://www.ogj.com/general-interest/economics-markets/article/14302481/global-oil-and-natural-gas-reserves-both-increase

Currently, the Western Desert contributes the largest share of oil production in Egypt (56%), followed by Gulf of Suez (23%), the Eastern Desert (12%) and Sinai (9%)[10]. Production activities are undertaken by over 40 local and international companies, including Apache, BP and Dragon Oil.[11]

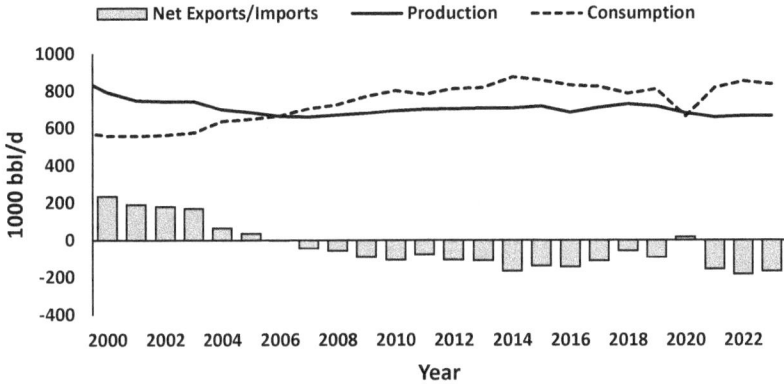

Figure 3.4: Egypt oil production, consumption, and trade (2000 to 2023)
Source: EIA, EI

The first natural gas field was discovered in 1967 in Abu Madi in the Nile delta. This was the start of natural gas discoveries in Egypt. In 1969 Abu Qir offshore natural gas field was discovered in the Mediterranean Sea, which was the first offshore natural gas field discovered in Egypt.[12] Several more discoveries followed onshore, including the Abu El Gharadig natural gas field in the Western Desert in 1971.

Despite numerous discoveries in Egypt prior to 2000, the Eastern Mediterranean region was not seen as a major natural gas province and most discoveries, onshore or in the shallow waters of the Nile Delta Basin were small. The interest in the region, which gained momentum following the discoveries of natural gas in Israel in 1999, then led to the further activities in Levantine Basin and the

---

[10] https://www.petroleum.gov.eg/en/gas-and-petrol/discovery-search-production/Pages/
Petroleum.aspx
[11] https://egyptoil-gas.com/maps/egypt-oil-and-gas-concession-map-february-2024/
[12] https://www.petroleum.gov.eg/en/gas-and-petrol/discovery-search-production/Pages/gas.aspx

adjacent basins.

In terms of natural gas, Egypt's proven reserves were 63 Tcf in 2023,[13] (a decline from a peak of 77.2 Tcf in 2011),[14] making it the fourth largest in Africa and ranking it 19th globally. As a result of the significant exploration activity continuing in this decade, further discoveries were made in 2022 and 2023, with more expected in the future,[15,16,17,18] including significant potential in the Red Sea.[19]

The rank of Egypt's reserves, relative to other African countries, has been slipping in the last few years. There have been large discoveries in Mozambique,[20] Tanzania[21] and Senegal.[22] There are also exploration activities in South Africa.[23] In practice, as new discoveries volumes take a long time to be listed as official reserves and adopted by publications, this is not currently reflected in many reported rankings. It is likely that further discoveries in Eastern and Southern Africa will push Egypt's rank down the list in the future.

Figure 3.5 shows Egypt's annual natural gas production and consumption, demonstrating the changeable export-imports dynamics in the country. Note that the majority of the Egyptian natural gas fields are small and are often developed in clusters or as tie-backs to improve their economic feasibility. Currently,

---

[13] https://www.ogj.com/general-interest/economics-markets/article/14302481/global-oil-and-natural-gas-reserves-both-increase

[14] BGR Energiestudie 2023 - Daten und Entwicklungen der deutschen und globalen Energieversorgung, potential additional resources 443 Tcf.

[15] https://www.rigzone.com/news/egypt_makes_53_new_oil_and_gas_discoveries_in_2022-30-dec-2022-171572-article/

[16] https://www.upstreamonline.com/exploration/chevron-discovers-3-5-tcf-of-gas-offshore-egypt/2-1-1375482

[17] https://www.eni.com/en-IT/media/press-release/2023/01/eni-announces-a-new-gas-discovery-in-the-eastern-mediterranean-sea-offshore-egypt.html

[18] https://www.offshore-energy.biz/shell-finds-gas-in-mediterranean-sea-with-stena-drillship/

[19] https://dailynewsegypt.com/2019/02/17/new-study-indicates-high-potential-for-gas-discoveries-in-red-sea/

[20] Estimated recoverable reserves exceed 160 Tcf; https://ametrade.org/newsroom/mozambiques-natural-gas-resources-and-the-quest-for-economic-development/.

[21] https://academic.oup.com/book/40396/chapter/347209049

[22] https://energycapitalpower.com/developments-senegalese-oil-gas-q2-2023/

[23] https://www.upstreamonline.com/exploration/supermajors-circle-as-scale-of-key-orange-basin-block-s-oil-and-gas-riches-revealed/2-1-1416677?%E2%80%A6

Mediterranean Sea fields contribute the largest share from natural gas production in Egypt with 62%, followed by Nile Delta with 19%, then Western Desert at 18%. Production activities are undertaken by over 20 companies, both local and international, including Eni, Apache, Shell and BP.[24] Other major international companies including Chevron, ExxonMobil and TotalEnergies have obtained exploration licences also.[25]

Egypt was a founding member of the Gas Exporting Countries Forum (GECF), which aspires to be the natural gas equivalent of OPEC, and the East Mediterranean Gas Forum (EMGF) – a regional organisation that promotes and develops natural gas reserves in the Eastern Mediterranean.

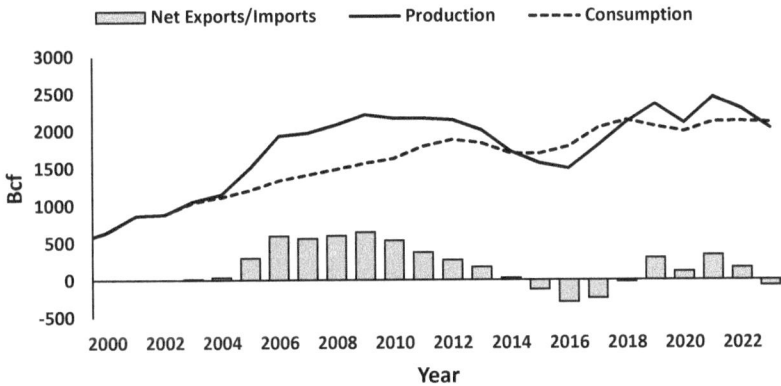

Figure 3.5: Egypt natural gas production, consumption and trade (2000 to 2023)

Source: EIA, EI

Egypt's initial aspirations pursued a dual policy of exporting natural gas via pipelines and LNG terminals. However, as discussed below, in the past it overstretched its resources, committing to export more than it can produce, and therefore often could not fulfil all its obligations. A classic case of overpromise and underdeliver.

---

[24] https://www.petroleum.gov.eg/en/gas-and-petrol/discovery-search-production/Pages/gas.aspx
[25] https://egyptoil-gas.com/maps/egypt-oil-and-gas-concession-map-february-2024/

Egypt is home to two liquified natural gas (LNG) export terminals. They are located on its northern shore, which are, to date, the only operational LNG liquefaction facilities in the Eastern Mediterranean region.[26] These are:

1. Damietta SEGAS LNG terminal, opened in 2005, with a capacity of 5 mtpa (i.e. million tons of natural gas per year). It has one train. which operates intermittently.

2. Idku LNG terminal, opened in 2005, with a capacity of 7.2 mtpa. It has two trains. which also operate intermittently.

These two LNG terminals have operated intermittently since opening. Both were often under-utilised, operated below capacity, and, even prior to Egypt turning into net natural gas importer in 2015, had to cease operating several times due to lack of available natural gas supply for export. This erratic operation led to successful compensation claims being awarded to operators, against the Egyptian government. Furthermore, expansion plans originally envisaged for both terminals had to be cancelled due to the lack of natural gas reserves.

In 2019, proposals were made to construct an LNG liquefaction terminal at Sinai, on the Red Sea, to utilise Israeli natural gas for export to Asia.[27] Limited progress was made on this proposal and the project is considered shelved.

Domestically, Egypt has a cross-country natural gas pipeline network/grid that connects its producing fields with the main population and its industrial centres. It is also expanding its local distribution system to supply residential and commercial users.

Currently, there are several cross-border natural gas pipelines connecting Egypt to its neighbours. Unlike the majority of natural gas pipelines, these pipelines have interesting stories and a controversial history. These pipelines are listed below:

---

[26] Libya has a mouthballed LNG liquefaction terminal.
[27] https://www.gasstrategies.com/information-services/gas-matters-today/egypt-and-israel-mull-red-sea-lng-export-facility-sinai

1. Arab Gas Pipeline (AGP),[28,29] initially constructed as a major export route, from Egypt, to send natural gas to Jordan, Syria and Lebanon. The project owners had grandiose plans for the pipeline, agreeing to extend it later to Turkey (see Section 3.5 – for details on Turkey's pipelines), with proposals to also link it to Iraq (see Section 4.3 – for details on Iraq's pipelines) and possibly Cyprus (see Section 3.4 – for details on Cyprus' pipelines). The 1200 km pipeline was constructed in phases. The first phase, commissioned in 2003, runs mostly onshore from Arish in Egypt to Aqaba in Jordan, with a small offshore section in the Gulf of Aqaba. The second phase, running across Jordan from Aqaba to Rehab was completed in 2005. The third phase, from Rehab in Jordan to Homs in Syria, was completed in 2008. The connections to Tripoli in Lebanon and to Banyas on the Syrian coast, were commissioned in 2009. The fourth phase, from Homs to Aleppo in Syria was not commissioned as a result of the Syrian civil war.[30] The interconnection, between Aleppo in Syria and Kilis in Turkey, was completed in 2011 but was never operational.

In the following years, several sections of this pipeline were reversed, starting with the offshore segment between Aqaba in Jordan and Taba, in Egypt sending natural gas from Jordan to Egypt (between 2015 and 2018), which was sourced from Jordan's floating storage regasification unit (FSRU) LNG import terminal. This was followed by the Jordanian section from the Israeli link point in northern Jordan to Aqaba, reversing in 2020 inside Jordan, then again in 2022 to send Israeli natural gas to Egypt.

Note that the stalled 2022 deal to send Egyptian natural gas

---

[28] https://www.gem.wiki/Arab_Gas_Pipeline
[29] Pipeline diameter is 36"; annual capacity is 10.3 bcm (360 Bcf), which can be doubled with additional compression stations.
[30] With this section not commissioned, natural gas between Homs and Aleppo can be transported using internal Syrian natural gas transmission system
https://www.arabfund.org/Default.aspx?pageId=470

to Lebanon was supposed to be using this pipeline (if it ever materialises), but the pipeline flow will not be reversed again. Instead, it will be an effective 'swap' deal – Egypt will not be sending its actual natural gas molecules to Lebanon– since the AGP pipeline sections are already reversed – instead, Israeli natural gas molecules will be supplied to Lebanon via Jordan and Jordan will get Egyptian natural gas in return.

2. Arish-Ashkelon pipeline[31,32] was originally constructed to export natural gas from Egypt to Israel. It is now effectively a branch of AGP, as it starts from the same take-off at Arish and runs for 90 km mostly offshore to Ashkelon in Israel. The pipeline was commissioned in 2008. The flow of the line was reversed in 2020, when it started exporting Israeli natural gas to Egypt. An upgrade to the pipe in southern Israel was completed in 2024 to ease bottlenecks and expand export capacity.

An extension from Ashkelon in Israel to Ceyhan in Turkey was once floated in early the 2000s, but the project never progressed.[33]

Other pipelines that are planned, proposed or cancelled include:

1. Israel-Egypt offshore pipeline to LNG facilities in Egypt (see Section 3.2 – for details on Israel's pipelines).

2. Israel-Egypt onshore pipeline linking to AGP (see Section 3.2 – for details on Israel's pipelines).

3. Gaza-Egypt pipeline, a proposed short offshore pipeline linking the Egyptian grid to Gaza Strip, to export potential Palestinian natural gas to Egypt.[34]

---

[31] https://www.gem.wiki/Arish%E2%80%93Ashkelon_Pipeline
[32] Pipeline diameter is 26"; annual capacity is 5-7 bcm (176-247 Bcf), which can be doubled with additional compression stations.
[33] https://www.annaharar.com/arabic/economy/oil/24082020
[34] https://www.africa-confidential.com/article/id/14860/Israeli_gas_enters_Sisi%27s_Gaza_calculations

4. Cyprus-Egypt offshore pipeline to connect to Egyptian LNG facilities (see Section 3.4 – for details on Cyprus' pipelines).

5. Egypt-Libya onshore pipeline, which was agreed in principle in 1997 and then again in 2002, to construct a pipeline linking the Egyptian natural gas network to the Greenstream pipeline in Libya (see Section 3.10 – for details on Libya's pipelines).[35] It was to utilise the Libyan internal transmission network, using the Libyan coast pipeline[36,37,38] and the link to Italy. It never progressed beyond announcements and was later abandoned.

6. Alternative EastMed pipeline,[39] proposed in 2021 as a technically and politically more favourable solution. The route is to run onshore along Egyptian coast, then an offshore section from Egypt to Greece, directly bypassing Cyprus.

7. Alternative EastMed Pipeline, a different route to construct an onshore pipeline from Egypt to Libya, then offshore to Greece or Italy. From an engineering point of view, it is less technically challenging as it avoids major geologic faults and runs in shallower water. (See Section 3.10 – for details on Libya's pipelines).

8. Libya-Egypt pipeline, linking Libya to Egyptian LNG facilities (See Section 3.10 – for details on Libya's pipelines).

As can be seen in Figure 3.5 above, the trajectory of Egypt's natural gas production tracks closely with the timeline of natural gas discoveries in Egypt. Dozens of discoveries were made by the

---

[35] https://www.projectfinance.law/publications/2004/june/libya-poised-to-be-a-major-gas-exporter/
[36] https://www.gem.wiki/Brega-Khoms_gas_pipeline
[37] https://www.gem.wiki/Khoms-Tripoli_gas_pipeline
[38] https://www.gem.wiki/Libya_Coastal_Gas_Pipeline
[39] https://www.euractiv.com/section/energy/news/athens-and-cairo-mull-changing-the-route-of-eastmed-pipeline/

year 2000 that made Egypt self-sufficient and later transformed Egypt into net natural gas exporter. Egypt's natural gas production reached a peak in 2009 at 2.2 Tcf. Then production started a sharp decline, which coincided with a rapid increase in demand for power generation, industry demand, residential and commercial consumption. It peaked again in 2021 at 2.5 Tcf, but has declined since. Egypt's power sector converted to natural gas, raising its share progressively to 79% in 2011 and 86% in 2019, before declining slightly when the lack of natural gas forced some power generation to revert back to heavy oil.[40] Note that domestic prices were and remain subsidised, with different prices (below market price) charged per end usage, while the government paid suppliers fixed prices.

Egypt started its pipeline exports to Jordan in 2003, and to Israel in 2008. It also started LNG exports in 2005. During the time of President Mubarak, Egypt overstretched its energy exporting targets, overcommitting on natural gas, beyond its means and what it could physically deliver.

There was massive controversy relating to the pricing of natural gas exported to Israel. The trading company buying the natural gas was paying the Egyptian government circa 1.5 US$/MMBtu, while Israel paid between 4–5 US$/MMBtu from 2009-2011, for its pipeline for natural gas imports from Egypt. The difference was pocketed by middlemen, who were later convicted on corruption charges. At the time, Egypt became an effective Israeli energy colony, selling natural gas at that low price.

At first, Jordan paid circa 2-3 US$/MMBtu for imported Egyptian natural gas. The price initially agreed was too low – albeit higher than the price trading middlemen supplying Israel paid, but subsequent agreements consistently increased the price to reach around 5 US$/MMBtu.[41] For Egypt, the absurdity of the situation became clear when, looking at the different prices between Egypt's natural gas imports and exports, the country paid almost double

---

[40] BP/EI Statistical Review of World Energy
[41] The Egyptian agreement with Israel supplied the gas at cheaper rates.

that price for its imports.[42]

In 2011, Egypt began favouring its domestic market over its export customers. It started slowly cutting off most of its pipeline exports, with supply to Israel was halted in 2012, then again effectively in 2013 (although a partial intermittent operation occurred in 2014). It also stopped supplying Jordan. The natural gas pipelines from Egypt to Jordan and to Israel were under-utilized and since that time they suffered numerous terrorist attacks causing significant damage, causing repeated disruption.

Following the cancellation of the supply contract, Israel won a multi-billion US dollar judgment in international courts,[43] and then reached a settlement with Egypt to be compensated for the unilateral cancellation.[44] Furthermore, the Spanish operator of Damietta LNG terminal won a multi-billion US dollar compensation from Egypt for unfulfilled shipments.[45] Jordan (due to political reasons) did not pursue compensation claims, despite its entitlement to do so.

In what appears now to be a repeat regional story, in 2015, unable to meet its demands, some of Egypt's power generation was forced back to using fuel oil and natural gas share declined to 71%.[46] Egypt turned into a net natural gas importer. Initially, in a reversal of the roles between the two countries, it imported excess LNG from Jordan in 2015. It rapidly increased its LNG imports from Spring 2015, by constructing two LNG import terminals, contracting two FSRUs with a capacity of 4.2 and 5.7 mtpa respectively. These FSRUs were stationed at Ain Sokhna in the Red Sea between 2015 and 2018. Subsequently one left and the other was relocated to SUMED port in 2017. The remaining FSRU left late in 2023, leaving the country without LNG regasification capability.

---

[42] http://platformlondon.org/2013/04/01/egypt-to-keep-exporting-gas-to-jordan-despite-energy-crisis-searching-for-imports/
[43] https://www.newarab.com/news/egypt-pay-israeli-firms-3bn-after-losing-appeal
[44] https://www.middleeastmonitor.com/20170304-218178/
[45] https://energyegypt.net/egypt-to-pay-union-fenosa-2-billion-in-damietta-lng-dispute/
[46] Egypt uses natural gas also in many industries.

When Egypt was a natural gas importer, Israel thought it could start exporting natural gas there. Instead, because of significant new discoveries in the Zohr field in 2015, with projections of growing domestic supply from existing fields, plans to import Israeli natural gas went on the back burner. With the rapid development of Zohr field, which came online in 2017, Egypt significantly increased domestic production and effectively ended LNG imports by 2021 (although a handful of cargo deliveries continued till 2023). As production rapidly increased, it resumed the position of a net natural gas exporter, supported by further developments in Deep Nile Delta Basin and other offshore developments.

Feeling bullish with its increased production capacity, Egypt did not appear to learn from its previous mistakes and unfortunately repeated its previous behavior. Thus, although it used to be the main natural gas supplier to Jordan, a return to this position was highly unlikely. Although Egypt resumed limited natural gas exports to Jordan between 2018 and 2021, these gradually stopped once Israel started supplying Jordan in 2020.

Also, Egypt resumed its LNG exports gradually from 2016 and in 2022. These achieved a record export volume following the Russian invasion of Ukraine and Europe's scramble to find alternative natural gas supply to substitute the suspended Russian natural gas supply.

However, despite this possibly advantageous situation, Egypt's natural gas demand was increasing significantly also.

Although not fully publicised to Egyptian public, in 2018, Israel finally achieved its goal becoming a natural gas exporter to Egypt. It signed an agreement to supply Egypt for 15 years, using the Arish-Ashkelon pipeline with its flow reversed.[47] The Israeli natural gas is used to fulfil Egyptian domestic demand and to supply the LNG liquefaction terminals. The deal was agreed at a

---

[47] https://www.reuters.com/article/us-israel-egypt-natgas-idUSKCN1G31BK

price of circa 6-6.5 US$/MMBtu[48]. When the natural gas price spiked late in 2021, continuing throughout 2022, Egypt made a neat profit, bagging the difference between this price and LNG export price.

Then later in 2022, Israel and Egypt signed an agreement to formally supply the EU with Israeli natural gas, using the Egyptian LNG export terminals - a first official agreement for Israel supplying the EU. The agreement is short term as the EU is refusing to commit to long natural gas supply agreements in their attempts to endorse energy transition goals and phase out its natural gas consumption. With this additional natural gas volume, the Jordanian section of AGP flow was reversed, south of the Israeli link to the pipeline in northern Jordan, to export natural gas via Jordan to Egypt.[49,50]

In 2023 we appear to be in another déjà vu situation since the growth of natural gas domestic consumption in Egypt continues to rise and is exceeding the growth in local supply, hence eating from Egypt's natural gas export volumes. At the same time, mismanagement of Zohr fields and disappointing production level from other fields meant the production declined more than anticipated. Thus, to try to make more volumes available, the Egyptian government is discouraging local consumption and has raised the prices significantly for local users in 2022. It has shifted its power stations from natural gas to heavy oil. It has inflicted power cuts and blackouts throughout the country, for several hours a day, to curtail natural gas consumption and provide volumes for export, where LNG exports were briefly halted.[51,52] Stability issues in neighbouring Israel, due to the war in Gaza in 2023, halted deliveries of natural gas to Egypt for several weeks, forcing the Egyptian government to import its first LNG cargo in years.

---

[48] https://www.linkedin.com/pulse/egypt-energy-hub-myth-reality-ahmed-shams-el-din/
[49] https://www.spglobal.com/commodityinsights/en/market-insights/latest-news/natural-gas/021722-israel-approves-new-route-for-gas-exports-to-egypt-via-jordan
[50] https://www.gov.il/en/departments/news/ng_160222
[51] https://www.thenationalnews.com/mena/egypt/2023/11/01/power-cuts-make-an-unwelcome-comeback-in-egypt-as-high-temperatures-linger/
[52] https://www.offshore-technology.com/news/egypt-to-resume-lng-exports/

While the suspension of natural gas deliveries was short-lived,[53] it highlighted the vulnerability of natural gas supply in Egypt and the risks of relying on limited sources.

Things turned worse in 2024, when Egypt returned to LNG imports, with production declining and Zohr field reserves being downgraded.[54,55] But having already disposed of all its LNG regasification terminals few months earlier, Egypt had to initially rely on using Jordan's FSRU to secure its supplies, until in May 2024, it contracted a new ship at Ain Sokhna to be stationed there for 19-20 months.[56] Who knows what the future will bring?

Throughout this chaotic journey, there were several false starts and the short-lived excitements of another possible mega-discovery that might be larger than Zohr discovery, but the buzz often faded quickly. One example is the Narjis discovery which created great excitement,[57] but the actual volumes proved to be disappointing. Who knows, there may be another Zohr-size discovery soon (prior to downgrading it hopefully for Egypt, history will repeat itself.

Egypt is trying to be the 'jack of all trades', the hub of all hubs. It wants to be a crude oil hub, drumming up SUMED oil pipeline[58] and the vital position of Suez Canal for tanker traffic; it aspires to be a natural gas hub as discussed above; and it also wants to be a global logistics hub. The Egyptian authorities work to promote the potential of the country but alas with limited success, as endemic corruption, legal instability and security issues hinder these efforts. In addition, the 2023-present war in Gaza illustrated how

---

[53] https://www.agbi.com/article/israeli-gas-exports-to-egypt-resume-in-small-volumes/

[54] https://www.mees.com/2024/4/12/oil-gas/eni-egypt-more-zohr-woe-as-reserves-downgraded/a82b58f0-f8cd-11ee-94bd-bdbdb199e957

[55] https://32980588.isolation.zscaler.com/profile/e5d06480-472e-42f7-9101-0b4ab3763316/zia-session/?controls_id=096b1f60-3f4f-4f05-90fd-538eb5ad19b3&region=lon&tenant=0f9d9e0bcf57&user=8fea5ecd338f3978af1de858e82aac57c66384978b5f009f29916d53e26992c7&original_url=https%3A%2F%2Fjfdi.expert%2Fnews%2F903-en%2F&key=sh-1&hmac=2b5378dd13748d8e554fdcb264351f97e130dac7e7c1db4dfeaa3efb76f02e85

[56] https://hoeghlng.com/investor-relations/news/news-details/2024/Hegh-LNG-announces-agreement-to-deploy-FSRU-Hoegh-Galleon-to-Egypt/default.aspx

[57] https://www.petroleumafrica.com/mega-gas-discovery-for-chevron-offshore-egypt/

[58] The Sumed Pipeline (Suez-Mediterranean Pipeline) is an oil pipeline in Egypt, running from the Ain Sokhna in the Gulf of Suez to offshore Sidi Kerir port, Alexandria, on the Mediterranean Sea. It provides an alternative to the Suez Canal for transporting oil from the Persian/Arabian Gulf region to the Mediterranean.

vulnerable the position of Egypt in its attempts to be a 'hub'. A few missiles from the Houthis in Yemen have managed to disrupt trade routes in the Red Sea and exposed the many challenges to holding the hub position in this region.

Politically, Egypt is Machiavellian in its attempts to position itself as the regional natural gas hub. It secretly sabotages other natural gas pipeline export projects, while pretending to support them publicly. An example of this practice is the EastMed pipeline, which is in direct competition to Egypt's hub-positioning ambitions. It negotiated bilaterally with Cyprus to build an alternative pipeline that will effectively kill the project, while suggesting an alternative route, via Libya, to scupper the project. Its interference in Libyan internal affairs is a main factor in the continuous destabilisation of Libya, which serves Egypt's goals of derailing competing natural gas solutions in the region.

Commercially, it strengthened its ties with Israel and transformed itself into Israel's main natural gas customer. It positioned itself as key market in the short term, prompting Israel to look for alternative outlets to send its natural gas and therefore still considering floating liquefied natural gas (FLNG) options. Moreover, it is placing itself as the logical option to enable the commercialisation of Cypriot natural gas reserves, by posing as the most feasible option to allow these reserves to reach the global market. Still, all these manoeuvres are quite regional, and in reality, are insignificant on the global stage as the natural gas volumes involved are relatively small.

Even though Egypt agreed its maritime borders with Greece, it appears to be backtracking, following the rapprochement between Egypt, Saudi Arabia, UAE and Turkey in 2021, negotiating a better deal with Turkey. In this case Egypt will gain additional territories. In this region, geopolitical alignments often shift, appearing fluid and never static.[59]

Compared to other Eastern Mediterranean countries, Egypt

---

[59] https://www.inss.org.il/publication/turkey-egypt/

nowadays is only intermittently one of three net natural gas net exporters (the others are Israel and Libya), but unlike the other two, its 'sometimes' surplus position is marginal and shaky, so it is drifting quickly back into being a net importer. Its future role looks likely to be limited to a transit state that makes some money on transport and liquefaction fees, rather than gaining revenues from selling natural gas. Thus, Egypt's dreams of becoming the regional natural gas hub seems at best unlikely as the country lacks a source of additional natural gas - indigenous spare natural gas volumes to export, and it relies on rerouting production from other neighbouring countries, it is not the home to major regional pipelines and has no plans to construct such pipelines.

Quantitatively, Egypt is a regional natural gas heavyweight, accounting for 32% of the reserves, and 55% of the production, but it also accounts for a significant 41% of the consumption. Thus, its contribution to the export market is trivial and has in reality evaporated. As a source of natural gas, its significance continues to diminish. Based on past experience and performance, unless there are dramatic, far-reaching changes, it is questionable that Egypt will ever rise to the top in terms of natural gas in the region.

3.2   Israel

Since its establishment in 1948, Israel has been in constant conflict with its neighbours. It has fought several wars against the Arabs and now controls the historic Mandatory Palestine. It is an extremely militarised country; it is the only nuclear power in the region and is considered the strongest country in the region in military terms. Since 1977 several Arab countries have normalised their relationships with Israel, starting with Egypt, then Jordan, with Bahrain, UAE, Morocco and Sudan lately following suit. The continuing occupation of the West Bank and the destruction of the Gaza Strip mean that the on-going Palestinian-Israeli conflict remains a constant cause of instability in the region. Moreover, after decades of being close allies, Israel's relationship with Turkey has deteriorated considerably in the 2000s. This tension is one of the main factors in the current uneasy energy dynamics in the Eastern Mediterranean. Note - this section is written in the midst of the war of 2023-present between Israel and Hamas-led Gaza, which is expanding in the region, and whose outcome is still unknown and may lead to considerable changes in the region.

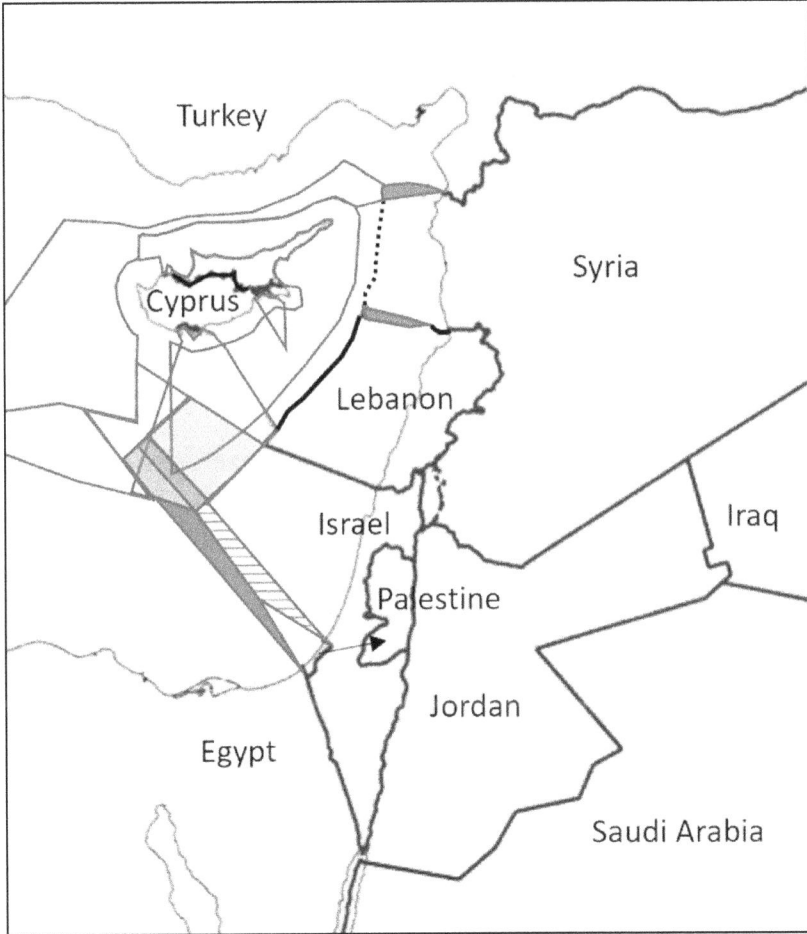

Figure 3.6: Levantine counties (Israel, Palestine, Syria and Lebanon) EEZs, highlighting disputed areas

Source: Drawn by author based on different sources.
Note 1: Borders are approximate and do not mean acceptance of claims.

Israel has relatively long coast on the Mediterranean Sea and a very short coast along the Red Sea. Israel's EEZ in the Eastern Mediterranean is currently governed by bilateral agreements or interim agreements. However, these arrangements are affected by other, conflicting, declarations and agreements between neighbouring countries. Note - Israel is not a signatory of UNCLOS,[60] thus it is not bound by the convention and the agreement. Figure 3.6 shows the current situation, which can be summarised as follows:

1. Israel-Cyprus EEZ is mostly covered by the bilateral agreement of 2010, based on equidistant or median line principles. However, many in Israel hold the view that this agreement is unfair, and that Israel should have gained more areas from Cyprus. At the moment, release from this agreement can only be achieved if Israel accepts the Turkish territorial claims against Cyprus, which is politically improbable.

2. In the hypothetical situation that Israel were to accept Turkish unilateral EEZ claims, with the potential expansion of Turkish territories, Israel would have maritime borders with Turkey and will gain additional hydrocarbon-rich areas from Cyprus.[61] If that were to happen, the UK could decide to claim EEZ rights for its sovereign bases in Cyprus, which could lead to a potential overlap dispute between Israel and the UK. At the moment, this seems politically unthinkable, but in the Middle East, anything is possible.

3. In 2022, the US brokered negotiations leading to the resolution of Israel and Lebanon's dispute over maritime borders and EEZ areas.

4. Part of the maritime border between Israel and the Palestinian Authority is governed by the interim Oslo agreement of 1994. To the north of this border, there is an

---

[60] United Nations Convention on the Law of the Sea entered into force in 1994 and was established to provide an overarching international agreement regulating the various uses of the world's oceans and seas.
[61] https://dayan.org/content/israel-turkeys-neighbor-across-sea-delimitation-maritime-jurisdiction-areas-between-turkey

overlap of claims between Israel and Palestine. In fact, Israel has de-facto control over both Palestine's maritime borders and EEZ. Thus far, a final settlement has not been reached. Note that the ongoing Israeli-Hamas war is further complicating this disputed border.

5. Depending on the outcome of the Israeli-Palestinian EEZ overlap claim, there may be a potential maritime border between Israel and Egypt. In practice, since Israel controls Palestine's maritime borders, this is the de-facto situation; however, a final settlement is yet to be agreed and, considering the status of the Palestinian-Israeli conflict, this may never be achieved or still a long way off.

Israel's EEZ claims in the Red Sea are more straightforward than the Eastern Mediterranean, with its maritime borders delimited with Egypt and Jordan.

Despite a long history of oil exploration in Israel, results are to date very disappointing. As of 2024, Israel has minimal crude oil reserves, and its production is trivial.[62] It currently imports almost all its oil needs. Note, Israel produced moderate crude oil quantities during its occupation of the Sinai oil fields during the years it occupied the area between 1967 and 1980.[63]

The story of Israel's natural gas exploration was similar to their oil story until 1999, when several discoveries happened. This has since transformed the regional energy landscape. With plenty of indigenous supply, Israel has diversified its primary energy usage significantly in the last two decades. It has moved steadily to rely on natural gas. After years of growth, the country's primary energy demand steadied over the last decade. Figure 3.7 shows the energy mix and the trends, illustrating the growing role of natural gas in the energy mix, at the expense of coal.

---

[62] This fact is sometimes challenged by stories in the Arab media that claim Israel is stealing huge amounts of Palestinian oil reserves. See for example https://www.middleeasteye.net/opinion/war-gaza-israel-brutal-drive-seize-palestinian-gas-reserves, or https://www.workers.org/2023/11/74864/. These are exaggerated stories as the quantities concerned are very small.
[63] Full evacuation of Sinai was completed in 1982.

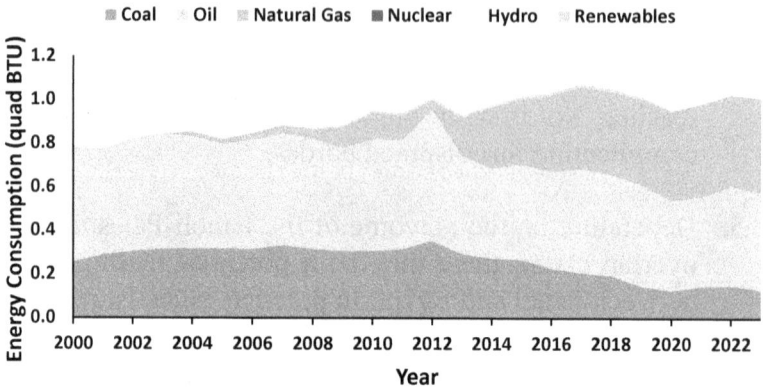

(a) Net consumption

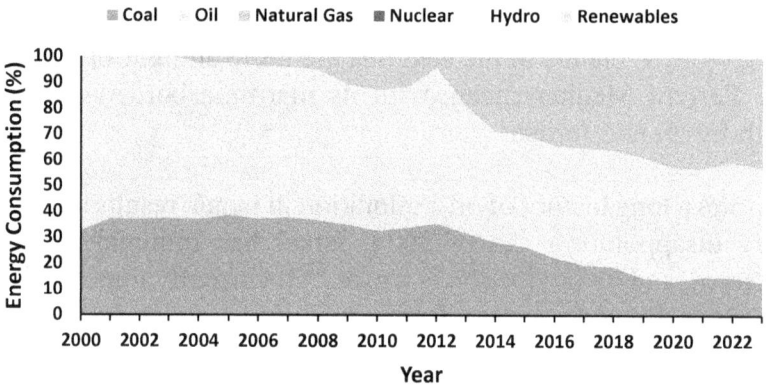

(b) Percentage consumption

Figure 3.7: Israels energy mix (2000 to 2023)

Source: EIA, EI

After years of limited success with minor onshore oil and natural gas discoveries, in 1999 Israel made a significant natural gas discovery in the Levantine Basin, with the Noa field offshore of Israel's south coast. Several findings were made in the following years including the Tamar natural gas field in 2009, followed shortly by Leviathan in 2010, with a combined volume exceeding 26 Tcf. Since then, even more fields have been discovered and exploration activities are booming in the Israeli territorial waters.

Published Israeli natural gas reserves at the end of 2023 stand at

6.216 Tcf, according to Oil & Gas Journal,[64] 7.1 Tcf according to BP/EI Statistical review,[65] with different numbers given by the US Energy Information Administration (EIA), the Federal Institute for Geosciences and Natural Resources (Bundesanstalt für Geowissenschaften und Rohstoffe - BGR) and other data sources. However, the reserve numbers published by these data providers underestimate the reserve values significantly. The actual declared reserves in Israel's largest field exceeds this and a simple field-by-field analysis gives an estimate of 43 Tcf.[66,67] This can satisfy Israel's needs for decades.

Historically, Israel has struggled to secure its oil and natural gas supplies. Following the Israeli withdrawal from Sinai, the Jewish State had to import its oil from as far away as Mexico, relying on Egypt for its natural gas requirements. Israel had many issues with these arrangements. As its own local natural gas resources were inadequate and needing to create security of supply, Israel had to look for other potential suppliers. Its relationship with Egypt was often tense and problematic, with the unpopular Israeli contract to import Egyptian natural gas being widely criticised in Egypt. Therefore, in order to secure its natural gas needs, Israel built a deepwater LNG regasification import terminal (1.8 mtpa),[68] which operated between 2012 and 2022, when its charter was terminated.[69] While this was not the best economic option, Israel was forced to adopt this as an interim solution.

The timing of the natural gas discoveries was ideal as it coincided with the decline in natural gas supplies to Israel from Egypt, when their domestic demand was rapidly increasing. This high domestic demand also forced the Egyptian government to eventually halt its

---

[64] https://www.ogj.com/general-interest/economics-markets/article/14302481/global-oil-and-natural-gas-reserves-both-increase
[65] The Energy Institute, based in London, is, as of 2023, the home of the Statistical Review of World Energy, published previously for more than 70 years by BP.
[66] BGR Energiestudie 2023 - Daten und Entwicklungen der deutschen und globalen Energieversorgung, potential additional resources 58 Tcf.
[67] To put it in context, Qatar's and Iran's reserves are 842Tcf and 1200 Tcf respectively.
[68] https://excelerateenergy.com/projects/hadera-deepwater-lng-terminal/
[69] https://www.argusmedia.com/en/news-and-insights/latest-market-news/2398860-excelsior-fsru-leaves-israel

natural gas exports to Jordan,[70] as well as reducing, then stopping, its LNG exports. The large discoveries in Israel allowed the country to become self-sufficient and subsequently turn the country into an exporter of natural gas.[71,72,73,74]

A significant proportion of the Israeli natural gas reserves are owned by the American company Chevron. They acquired Noble Energy in 2020, in partnership with the Israeli conglomerate Delek (partially via its subsidiary Delek Drilling (renamed NewMed in 2022)). This dominant ownership position caused antitrust concerns and led to restrictions being imposed on the two companies by the Israeli government. This forced them to divest some of their assets and prevented them from participating in further bidding rounds. Other fields are owned by London-based Greek-origin company Energean. As it stands, there is a very limited number of operators currently in Israel, although more companies participated in recent bidding rounds and obtained licences, including Eni, BP, Socar, Dana Petroleum, and Ratio Energies.

Figure 3.8 shows Israel's annual natural gas production and consumption, demonstrating the rapid growth in production, which is outpacing consumption in the country. Although Israel is not a member of GECF, it is a founding member of the EMGF.

At the moment, Israel's chosen export method for its natural gas is via natural gas pipelines (where part of the gas is liquified elsewhere), but consideration was given to establish LNG exporting capability, either onshore or offshore. After several

---

[70] Exports to Jordan were severely affected by the deteriorating security situation in Sinai and the gas pipeline linking Egypt and Jordan was subjected to numerous sabotage attacks.
[71] https://www.reuters.com/article/us-israel-natgas-leviathan-idUSKBN1YZ0H9
[72] https://www.spglobal.com/platts/en/market-insights/videos/market-movers-europe/011821-nord-stream-oil-energy-forum-iea-sustainability-week-french-nuclear
[73] https://www.offshore-mag.com/production/article/14074908/leviathan-gas-flowing-to-jordan
[74] Despite legal setbacks and adverse ruling by the Israeli Supreme Court concerning the original development deal, Leviathan project is progressing.
(http://www.wsj.com/articles/israel-supreme-court-rules-against-offshore-gas-deal-1459105458; http://www.rigzone.com/news/oil_gas/a/144787/Leviathan_Partners_Sign_3B_Israeli_Natgas_Supply_Deal)

abandoned proposals including Tamar floating LNG,[75,76] and Eilat LNG,[77,78,79,80] serious consideration is still being given to utilise floating LNG liquefaction offshore the Israeli coast, to export Leviathan natural gas reserves. This project is still in the planning stage and several proposals are underway, as discussed later in this section.

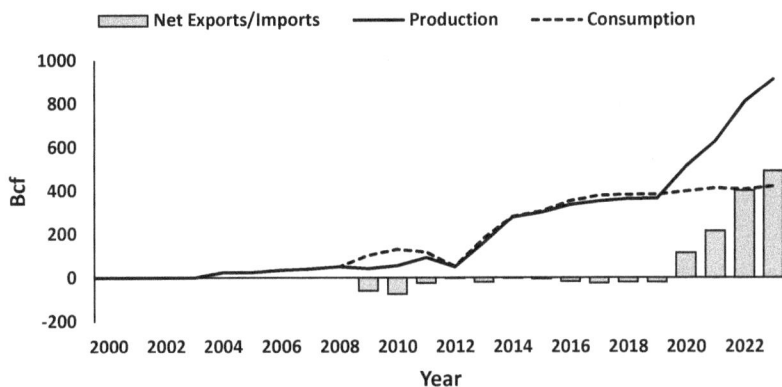

Figure 3.8: Israel natural gas production, consumption, and trade (2000 to 2023)

Source: EIA, EI

Domestically, Israel has a cross-country natural gas pipeline network/grid connecting producing fields with the main population and industrial centres.[81] It is also expanding its local distribution system to supply both residential and commercial users.[82]

Currently three cross-border natural gas pipelines connect Israel to its neighbours. These pipelines are listed below:

---

[75] https://2b1stconsulting.com/noble-awards-tamar-flng-feed-contract-to-pangea-lng/

[76] https://www.upstreamonline.com/weekly/noble-faces-flng-hurdle-at-israels-tamar-project/1-1-992799

[77] https://www.upstreamonline.com/lngunlimited/israeli-pair-enters-eilat-into-export-plant-race/1-1-988944

[78] http://abarrelfull.wikidot.com/eliat-lng-terminal-project

[79] https://www.gov.il/BlobFolder/reports/ng_commitee/he/INGL.pdf

[80] https://www.gem.wiki/Eilat_FLNG_Terminal

[81] https://www.ingl.co.il/?lang=en

[82] https://english.leumi.co.il/static-files/10/LeumiEnglish/Economic_Weekly/Natural_gas_Nov2021.pdf?WT.mc_id=in00002

1. Arish-Ashkelon pipeline is effectively a branch of AGP, linking Arish in Egypt and Ashkelon in Israel. (See Section 3.1 – for details on Egypt's pipelines). The offshore pipeline operated between 2008 and 2012, exporting natural gas from Egypt to Israel, then stopped for several years. It was subsequently reversed and, since 2020 has been used to export natural gas from Israel to Egypt. It was upgraded and extended in southern Israel in 2023-24 to remove bottlenecks.

2. Israel-Jordan interconnector (Northern Route),[83] is a 65 km onshore pipeline, that faced (and continues to face) strong public opposition in Jordan.[84] It was commissioned in 2020, linking the Israeli grid to AGP at Al-Khanasri, near Mafraq in Jordan and transports natural gas from Leviathan field to Jordan. Some of the natural gas quantities are then destined for other markets, currently Egypt and potentially to Lebanon via Syria.

3. Israel-Jordan interconnector (Southern Route), a low-key short, 17 km, onshore spur, that links the Israeli grid at Sdom, to Jordanian industrial facilities at Ghor Safi in the Southern Dead Sea area, to supply energy to the chemical industries complex.

Other pipelines are planned, proposed or cancelled. These include:

1. East Mediterranean Pipeline (EastMed),[85,86] is a planned 1900 km offshore/onshore natural gas pipeline, directly connecting natural gas resources in Israel and Cyprus to Europe, reaching mainland Greece, after passing Cyprus and Crete. It will then connect with other pipelines in Greece, allowing the natural gas to flow to Italy and

---

[83] Pipeline diameter is 36"; annual capacity is 10.3 bcm (360 Bcf),.
[84] https://www.gem.wiki/Israel_Jordan_Gas_Pipeline
[85] https://www.gem.wiki/EastMed_Gas_Pipeline
[86] Pipeline diameter varies between 24 to 48 inches, and proposed capacity is initially 10 bcm (353 Bcf), then doubling to 20 bcm (708 Bcf)

beyond.

The route has four main sections - one onshore and three offshore:

- 165 km offshore, connecting natural gas fields offshore of Israel to Cyprus.
- 732 offshore, between Cyprus to the east of Crete.
- 421 km offshore. from Crete to mainland Greece.
- 317 km onshore, across Western Greece

The project has been approved by the governments of three countries and is currently in the design stage. However, it is technically very difficult as it passes through active seismic regions and several fault lines. In addition, it reaches a depth of over 3000 m.

For several years, this pipeline was often viewed as the cornerstone of the regional natural gas policy. The EastMed Pipeline is the flagship project of the region that could unleash its potential, and unlock its resources, thus enabling integration with Europe and offering a viable solution to market the region's resources. It would also cement the region's essential position to the security of the supply of natural gas to Europe.

However, in reality, this project is unlikely to materialise as there are other alternatives for exporting natural gas from the region.

In February 2022, following the US announcement of no support, the EastMed project appeared to have been cancelled. Unexpectedly, following the Russian invasion of Ukraine in late Feb 2022, with Europe scrambling to find alternatives to substitute Russian natural gas, it appeared to have been given a reprieve. However subsequent proposals regarding LNG liquefaction projects in Israel and/or Cyprus, appear to have parked this project once again.

Note - the current EastMed route passes through the disputed EEZ of Turkey. Alternative routes for the project were floated several times, including routes via Egypt or Turkey, bypassing Cyprus or via Egypt and Libya, even bypassing Greece, directly to Italy. (See Section 3.1 – for details on Egypt's pipelines, Section 3.5 – for details on Turkey's pipelines and Section 3.10 – for details on Libya's pipelines).

2. Israel-Cyprus pipeline,[87,88] this is a proposed 215 km offshore pipeline to link the Tanin, Karish and Karish North fields to a proposed LNG export terminal (to be constructed at Vassiliko) in Cyprus. This project cannot be pursued as stand-alone or in conjunction with the EastMed pipeline and thus may be incorporated as a first phase of the EastMed project, if it ever materializes. In December 2022 Cyprus indicated that it prefers a cheaper, more flexible LNG option to export its natural gas, to the EastMed pipeline.[89] Importantly, as covered in Section 3.4, agreements regarding Cypriot natural gas developments have not been finalized yet.

3. Israel-Turkey pipeline, an offshore pipeline, which has been proposed several times, to export Israeli natural gas to Europe, via Turkey, linking at Ceyhan. Economically it is a viable alternative to and considerably cheaper than the EastMed pipeline. However, it faces several political obstacles,[90,91] despite reported behind-the-scenes negotiations between Israel and Turkey.[92]

4. Israel-Egypt offshore pipeline,[93,94] would link the

---

[87] https://www.gem.wiki/Israel_Cyprus_Gas_Pipeline
[88] Pipeline proposed capacity is 4 bcm (141 Bcf).
[89] https://www.timesofisrael.com/cyprus-weighing-proposed-pipeline-to-funnel-natural-gas-from-israel/
[90] https://www.mining.com/web/turkey-israel-gas-pipeline-rumors-are-geopolitically-tectonic/
[91] https://www.energy-reporters.com/policy/eastern-mediterranean-gas-testing-the-field/
[92] https://english.alarabiya.net/business/energy/2022/03/29/Israel-Turkey-gas-pipeline-an-option-for-Russia-wary-Europe-but-the-hurdles-are-many
[93] Pipeline proposed capacity is 10 bcm (353 Bcf).
[94] https://thearabweekly.com/egypt-agrees-israel-new-gas-pipeline-project

Leviathan field in Israel to Egypt, in order to supply the country's under-utilised LNG liquefaction facilities at Idku[95] and Damietta. The LNG would then be exported to Europe.

5. Israel-Egypt onshore pipeline,[96,97] a proposed link from the Israeli grid to Egypt, via Nitzana, connecting to the Arish-Aqaba segment of AGP. This was approved in 2023.[98]

6. Israel-Gaza pipeline,[99] which would link the Israeli grid to the Gaza Strip, it could potentially be used to transport Palestinian gas to Israel.[100] The ongoing war between Israel and Hamas-led Gaza has placed the future of this pipeline in considerable doubt.

7. Israel-West Bank pipeline,[101] a proposed short onshore pipeline linking Afula in Israel to Jenin in Northern West Bank. The outcome of 2023-present Gaza war will determine the future of this pipeline.

Rumors and conspiracy theorists often mention a potential Iraq-Israel natural gas pipeline without citing any references or providing evidence. This pipeline is not considered as practical nor is it feasible, so it is certainly a myth, mentioned in media outlets only to create controversy.

As seen in Figure 3.8, following earlier discoveries in Israel in 1999, the country gradually shifted away from using coal to natural gas. With further discoveries throughout 2000s, reserves increased significantly. These natural gas discoveries allowed Israel to become effectively self-sufficient since 2013.

---

[95] https://www.gem.wiki/Israel%E2%80%93Egypt_Offshore_Gas_Pipeline
[96] https://www.gem.wiki/Israel%E2%80%93Egypt_Onshore_Gas_Pipeline
[97] Pipeline proposed capacity is 3-5 bcm (106-176 Bcf).
[98] https://www.energy-sea.gov.il/home/news-publications/new-export-pipeline-to-egypt-approved-by-government/
[99] https://www.ingl.co.il/?lang=en
[100] https://www.africa-confidential.com/article/id/14860/Israeli_gas_enters_Sisi%27s_Gaza_calculations
[101] https://www.ingl.co.il/?lang=en

As mentioned earlier, the Israeli domestic power sector has converted from using dirtier hydrocarbons, i.e. from coal and oil to natural gas. The domestic sector is the largest consumer of natural gas in the country. Initially, Israel's local production (started in 2004[102]) did not supply its needs and it had to import natural gas to meet the demand. At that time, the country diversified its sources by importing natural gas from Egypt in 2008,[103] as well as constructing an LNG (Liquefied Natural Gas) import terminal, which started operation in 2012. Note that although those LNG facilities were initially important when they were covering a short period of supply weakness, their future role was supposed to supply periods of peak demand. Initially, it was planned that the terminal would be retained for security of supply, and thus would be maintained, as part of the mix, to cover annual maintenance or if there were unexpected disruptions to domestic natural gas production. However, with more fields developed in Israeli waters providing alternative supply sources, the terminal was discontinued in 2022, as mentioned earlier.

Exploration momentum is still active, and Israel continues to offer new exploration blocks[104] in successive bidding rounds. Despite all the discoveries, the Israeli government in an attempt to manage its domestic resources carefully, has been slow in liberalising its natural gas sector. Israeli law stipulates that 60% of Israel's reserves would be retained for the domestic market, with the remaining 40% available for export.[105,106] However, concerns about long-term export opportunities, as well as the desire to become a new supplier to EU to replace Russian gas, may lead to greater near-term exports. Due to this restriction, the size of reserves is proving to be a continuous issue between the government and the fields' operators, as it can limit the size of allowed exports and thus affects the fields' development plans.

---

[102] Production started from Mari-B field, which, together with the smaller Noa and Pinnacles fields, is known as the Yam Tethys project.
[103] Egyptian supplies peaked in 2010 at 310 MMscf/d but proved unreliable, and were ultimately stopped in April 2012.
[104] An oil exploration block is a large area that is awarded to oil drilling and exploration companies by a country's government.
[105] https://www.reuters.com/article/israel-natgas-idUSL5N0EZ0BQ20130623
[106] https://www.bloomberg.com/news/articles/2021-10-14/israel-says-it-can-t-export-more-gas-even-amid-soaring-demand?leadSource=uverify%20wall

Due to the substantial size of the Israeli natural gas reserves, Israeli demand alone would not justify further natural gas fields' developments. Accordingly, in order to justify further development, Israel needed to find export markets for its future natural gas production. The Tamar fields started production in 2013, and its production was allocated to the local market. While all natural gas from Tamar's first two phases had previously been allocated to domestic buyers on long-term contracts, it was clear that further development phases there, or in other fields, will need to find different markets for export.

Following several aborted plans to develop some of the fields as LNG projects, Israeli focus shifted to developing multiple regional natural gas pipeline links. However, following several legal hurdles due to the concentration of natural gas reserve ownership, these plans suffered initial delays. A legal settlement was reached between the operators and the government in 2015, paving the way for the development of the Leviathan field to go ahead. The field was developed in several phases, with the initial phase starting production in 2019.

Subsequently, natural gas from the Tamar field was agreed to be exported to Egypt and Jordan. Initial agreements signed by two Jordanian companies to import Israeli natural gas from Tamar field in 2014 were suspended and then renegotiated to reduce the price. In 2015 Israel approved two pipeline routes for the delivery of natural gas exports to Jordan.[107]

In the Leviathan field's case, during its initial development phase, secured contracts only covered a proportion of its production, initially to supply local Israeli clients, a 15-year supply contract to NEPCO of Jordan and to supply Egypt's domestic market, with its LNG terminals subsequently negotiated.

The 2017 natural gas deal (to supply Jordan) agreed prices of

---

[107] Israel approved a 15.5 km (9.6 mile) pipeline to Jordan in September 2015. The pipeline is an extension to the current Israeli grid.

exported natural gas at around 5-6 US$/MMBtu. This amount is significantly lower than the price quoted for the scrapped 2014 deal, which priced the natural gas at around 7.5-9 US$/MMBtu.[108] That deal included an allowance for the construction of the associated pipeline between Israel and Jordan. At the time, that figure was deemed reasonable as it was cheaper than LNG market prices being imported by Jordan, though it was more expensive than natural gas imported from Egypt.

In order to put the numbers in perspective, from 2009-2011, Israel paid circa 4–5 US$/MMBtu range for its natural gas imports from Egypt via pipeline, while its LNG imports costs were reported to be as high as 18-20 US$/MMBtu, (although costs then fell along with global spot prices). Likewise, Israel had to pay relatively higher prices when it ramped up its domestic production in 2010 and 2011, to compensate for faster depletion rates. Since 2014, its supply from Tamar averaged 5-6 US$/MMBtu, although the numbers dropped slightly in the early 2020s.

As discussed earlier in Section 3.1, in 2018, Israel reached an agreement to export natural gas to Egypt via the reversed Arish-Ashkelon pipeline that started in 2020.[109] This was followed by further agreement with the EU to import Israeli natural gas, in the form of LNG, utilising Egypt's LNG export terminals.

Following the settlement of the Israeli-Lebanese maritime dispute in 2022, natural gas from Karish field started flowing in 2022. It is destined to local markets and potential exports to Egypt.

While at the moment, Israeli natural gas exports rely on pipelines and other infrastructure in Egypt and Jordan, Israel prefers to rely on itself for export, either via FLNG,[110] or via a pipeline to Greece, rather than via pipeline to Turkey (with unfriendly relations) or via Egypt's facilities. However, for the time being, exporting via Egypt is the cheapest solution. Unfortunately, current export routes to Egypt are not sufficient and more pipelines are needed, as

---

[108] Much lower than the price it currently pays for crude oil in power generation.

[109] https://www.middleeastmonitor.com/20200115-israel-starts-gas-exports-through-egypt/

[110] https://www.upstreamonline.com/focus/lng-may-topple-east-med-pipeline/2-1-1375761

discussed earlier, either directly offshore to Egypt's LNG terminals or connecting to existing pipelines onshore.

Now, with further development of Leviathan planned,[111] an FLNG solution to export the gas remains actively but reluctantly on the agenda,[112,113] despite alternative plans of new direct pipelines to Egypt. Some even advocate using the proposed facility to export Cypriot natural gas, but the probability of this materialising is minute. The opposite though is being considered, with proposals to build a natural gas pipeline from Israeli natural gas fields to Cyprus and to then utilise a proposed liquefaction facility in Cyprus to export the natural gas.[114] Note that EastMed alternative is all but dead.

In fact, the EastMed gas pipeline is now being replaced by an electric interconnector. This new project, known as the EuroAsia Interconnector, will link the power grids of Israel, Cyprus and Greece.[115] It is an undersea electricity cable that will help integrate renewable energy sources and enhance energy security in the region. The EU is supporting the project and has allocated over 600 million Euros to it. This project is progressing and will significantly reduce Cyprus' energy isolation.[116]

The shift, from the EastMed natural gas pipeline to the EuroAsia Interconnector, signals a marked move in policies. This policy shift, where gas-to-power can be used as a way to transport natural gas from the region, integrating it with other renewable sources, aiming towards sustainable energy solutions and reducing environmental impact as part of energy transition map in the region.

---

[111] https://www.offshore-technology.com/data-insights/oil-gas-field-profile-leviathan-phase-1b-conventional-gas-field-israel/?cf-view
[112] https://lngprime.com/asia/chevron-newmed-energy-approve-budget-for-leviathan-flng/74397/
[113] https://www.mees.com/2023/11/24/oil-gas/leviathan-flng-plans-stall-as-costs-rise/8cafdd40-8ace-11ee-8b26-2f19f8a0c4aa
[114] https://maritime-executive.com/article/israel-and-cyprus-agree-on-landmark-lng-gas-export-project
[115] https://greekreporter.com/2022/01/27/energy-israel-cyprus-greece-electricity-link/
[116] https://www.offshore-energy.biz/euroasia-interconnector-entering-construction-phase-with-launch-date-set/

However, stability concerns in Israel due to the war in Gaza in 2023-present, halted deliveries of natural gas to Egypt for several weeks, affecting LNG exports. While the suspension of natural gas deliveries was short-lived, it highlighted the vulnerability of natural gas supply from Israel and the risks to the LNG exports in Egypt who rely on Israeli imports instead of Egyptian limited sources.

As it stands, Israel's natural gas production is expected be supplied to domestic or regional markets, with excess volumes commercialised through offshore or onshore liquefaction facilities, to be exported to Europe or beyond.

Relative to other Eastern Mediterranean countries, Israel is currently only one of three natural gas net exporters (the others are Egypt and Libya). In 2024 Israel is the largest regional exporter and its production is growing, unlike the other two. Thus, as a new regional energy 'super' power, Israel's aim is to leverage this power in order to be accepted in its local neighbourhood and integrate into the region. Israel has big dreams, but is still struggling to sell its natural gas. The ongoing Gaza war of 2023-present may yet derail many of the future plans. Despite continuing America, British and some EU support for Israel, nowadays, uncertainty prevails.

Quantitatively, Israel is now a regional natural gas heavy-weight, accounting for 19% of the reserves and 25% of the production, while it only accounts for 8% of the regional consumption. Thus, in natural gas dynamics terms, Israel's significance as a source of natural gas, continues to strengthen, with its influence in the regional energy market destined to grow further. However, its influence on the global markets will remain insignificant since the excess volumes to be exported outside the region will not make a dent on global markets.

3.3 Palestine

In this book, Palestine refers to the partially recognised Arab state in the West Bank and Gaza Strip – two areas still occupied by

Israel. Although the Palestinian National Authority (PNA) supposedly enjoyed nominal limited governance of the areas, in fact the PNA is militarily and economically under the control of Israel. The PNA only has limited control of areas of the West Bank, whereas Gaza Strip was controlled by the Islamic faction Hamas until the 2023-present war. Note this section is written in the midst of the war of 2023-present between Israel and Hamas-led Gaza, whose outcome is still unknown and may lead to considerable changes in the region.

Palestine is not a member of GECF, but it is a founding member of the EMGF.

The Palestinian territories have a short length of coastline along the Mediterranean Sea. The Palestinian Authority EEZ areas are yet to be finalised, due to the ongoing Palestinian-Israeli conflict. Figure 3.6 (in Section 3.2) shows the current situation which can be described as follows:

1. Part of the maritime border between Israel and the Palestinian Authority is governed by the 1994 interim Oslo agreement between Israel and the Palestinian Authority, northeast of this, there is an overlap of claims between Israel and Palestine.

2. Similarly, part of the maritime border between Egypt and the Palestinian Authority is governed by the 1994 interim Oslo agreement between Israel and the Palestinian Authority. There is an overlap of claims between Egypt and Palestine northwest of this also.

3. If either or both claims are solved in Palestine's favour, Palestine will have a maritime border with Cyprus, obviously, this is yet to be determined. In practice, since Israel controls Palestine's maritime borders, a final settlement is yet to be agreed and, considering the long-standing Palestinian-Israeli conflict and the current Israeli military actions, this may never be achieved.

4. Should the Palestinian Authority accepts Turkish unilateral EEZ claims and the potential expansion of them, Palestine would have maritime borders with Turkey and would gain areas from Cyprus.[117, 118] In this scenario, if the UK decides to claim EEZ rights for its sovereign bases in Cyprus, there is a potential overlap dispute between Palestine and the UK. This is politically not being considered by the British government.

Palestine, like Israel, despite many years of exploration, had discovered minimal oil reserves in its territories. It imports all its oil supplies from Israel. The situation regarding natural gas discoveries was also similar to Israel, until 2000, when the first field in Gaza's waters was discovered. There are numerous conspiracy theorists dispute this and believe that the governments in the region are hiding information regarding the size of the Palestinian reserves or that even the latest war in Gaza in 2023-present is because of natural gas.[119, 120] See the discussion later in the section.

Primary energy demand in the Palestinian Authority is dominated by oil (see Figure 3.9). Currently, the Palestinian Authority imports all its energy needs from Israel both as primary energy and as electricity. It has no natural gas demand. Although there were plans to utilise natural gas as the chosen fuel for power generation, Israel has not given permission to develop natural gas discoveries yet. As long as Israel prevents the development of indigenous natural gas resources, the PNA will have to continue to import Israeli natural gas.[121]

After the first field, Gaza Marine, was discovered in 2000, further exploration identified a second field, but exploration activities have been halted since then due to restrictions imposed by Israel.

---

[117] https://uwidata.com/18631-turkish-palestine-eez-delimitation-proposal-causes-panic-in-israel-and-greece/
[118] https://www.alestiklal.net/en/view/8965/this-is-how-turkey-demarcates-the-mediterranean-common-boundaries-with-palestine
[119] https://www.noonpost.com/27373/
[120] https://www.alestiklal.net/ar/article/dep-news-1615536897
[121] In 2015, an agreement to supply Israeli gas to the PNA for power generation was cancelled.

At the start of 2021, Palestine's natural gas reserves stood at 1.4 Tcf.[122,123,124] Interestingly, the Oil & Gas Journal does not report any reserves for Palestine. Note - the matter of energy for Israel and Palestine is difficult to untangle. In many cases, it is hard to find separate or independent numbers for Palestine and often these numbers are included within Israel's data.

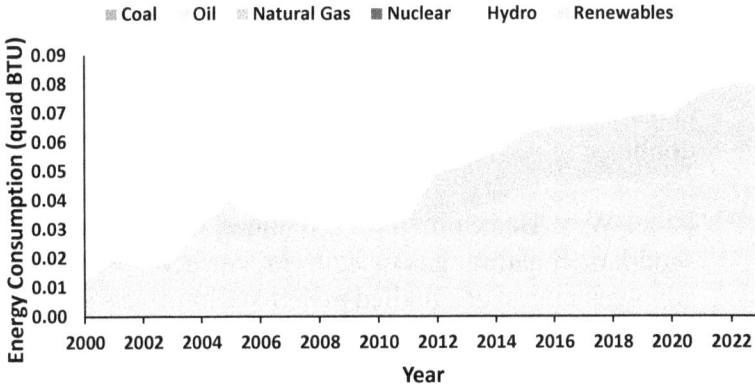

(a) Net consumption

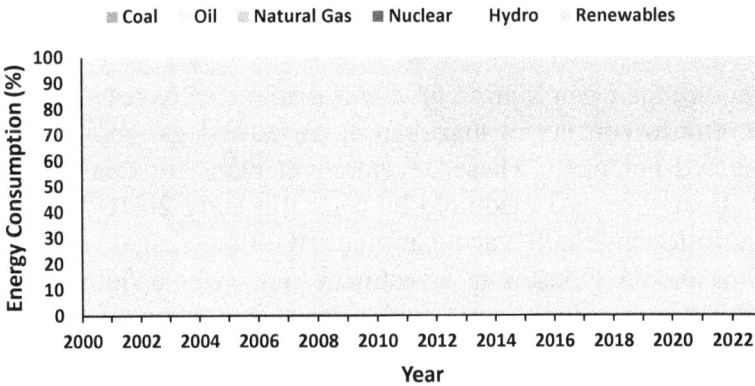

(b) Percentage consumption

Figure 3.9: Palestine energy mix (2000 to 2023)

Source: EIA

[122] http://www.palestine-studies.org/jps/fulltext/162608
[123] BGR Energiestudie 2023 - Daten und Entwicklungen der deutschen und globalen Energieversorgung, potential additional resources 13.3 Tcf.
[124] https://unctad.org/system/files/official-document/gdsapp2019d1_en.pdf

There are currently no natural gas pipelines connecting the PNA to its neighbours. However, there are several proposed pipelines. These are:

1. Israel-Gaza pipeline, a planned onshore pipeline to send Israeli natural gas to Gaza Strip. (See Section 3.2 – for details on Israel's pipelines). If Palestinian natural gas exports were ever to happen, this pipeline could be reversed. The ongoing war between Israel and Hamas-led Gaza has placed the future of this pipeline in considerable doubt.

2. Israel-West Bank pipeline, a planned onshore pipeline to send Israeli natural gas to Jenin in Northern West Bank, to supply a natural gas fuelled power station. (See Section 3.2 – for details on Israel's pipelines) The outcome of 2023-present Gaza war will determine the future of this pipeline.

3. Gaza-Egypt pipeline – as described earlier. (See Section 3.1 – for details on Egypt's pipelines).

Although the Gaza Marine field was discovered over two decades ago, due to conflict in that region, its natural gas reserves have remained untapped. These development plans for this gas field, first floated by BG (acquired by Shell) in early 2000s, [125, 126] and then, following Shell's acquisition of BG, floated again by CC Oil & Gas and the Palestinian Investment Fund (who acquired Shell's working interest in the asset, with each party now owning 50%, [127]) have stalled for years.

After years of inactivity, in 2022 news broke that Egypt was considering plans to develop them, [128,] and initial agreements were

[125] https://www.spglobal.com/commodityinsights/ko/market-insights/latest-news/natural-gas/031413-update-bg-group-in-secret-talks-to-develop-gaza-marine-gas-field-report
[126] https://electronicintifada.net/content/gaza-siege-intensified-after-collapse-natural-gas-deal/7312
[127] https://www.ccenergyltd.com/operations/palestine/overview#:~:text=The%20Gaza%20Marine%2D1%20(GM,up%20to%2037%20mmscf%2Fd.
[128] https://www.reuters.com/business/energy/israel-gives-nod-gaza-marine-gas-development-wants-security-assurances-2023-06-18/

announced in 2023,[129] shortly before the latest war. Needless to say, this inevitably put these plans to hold again. Ironically it was suggested that Israel had approved these plans, following an assessment that the economic benefits from this project would persuade Hamas to desist from any military action.[130]

The subject of Palestinian natural gas reserves is a prominent internation media topic. There is significant misinformation, propaganda, biased opinions and unfounded claims. Discussing all of them would require another book but it is important to have an awareness of a selection of them. Below I provide a small selection, just to give a flavour of the debate printed in the Middle East as well as discussed heatedly in both traditional and social media.

Let's look at the stories regarding Jordan's intent to import Palestinian natural gas. Between 2014-2015, articles appeared in Jordanian and Arab media, claiming that Jordan would import Palestinian natural gas. [131, 132, 133] Despite announcements made by Jordanian officials concerning agreements made to import as much as 180 MMscf/d of Palestinian natural gas to Jordan, in the end, this never happened. In fact, it is highly unlikely that this natural gas supply will occur in the future. Firstly, the required resources had not yet been developed, secondly, the PNA would need to prioritise its supplying its own demand, so it would be more likely to consider fulfilling its domestic consumption before considering

---

[129] https://www.newarab.com/news/exclusive-hamas-allows-development-gas-field-gaza

[130] https://www.africa-confidential.com/article/id/14860/Israeli_gas_enters_Sisi %27s_Gaza_calculations

[131] https://www.alaraby.co.uk/%D8%A7%D9%84%D8%A3%D8%B1%D8%AF%D9%86-%D9% 8A%D8%A8%D8%B1%D9%85-%D8%A7%D8%AA%D9%81%D8%A7%D9%82%D8%A7% D9%8B-%D9%84%D8%B4%D8%B1%D8%A7%D8%A1-%D8%A7%D9%84%D8%BA%D8 %A7%D8%B2-%D8%A7%D9%84%D9%81%D9%84%D8%B3%D8%B7%D9%8A%D9%86% D9%8A-%D9%85%D9%86-%D8%BA%D8%B2%D8%A9

[132] https://www.alaraby.co.uk/%D8%A7%D9%84%D8%A3%D8%B1%D8%AF%D9%86-%D9% 8A%D8%A8%D8%B1%D9%85-%D8%A7%D8%AA%D9%81%D8%A7%D9%82%D8%A7% D9%8B-%D9%84%D8%B4%D8%B1%D8%A7%D8%A1-%D8%A7%D9%84%D8%BA%D8 %A7%D8%B2-%D8%A7%D9%84%D9%81%D9%84%D8%B3%D8%B7%D9%8A%D9%86% D9%8A-%D9%85%D9%86-%D8%BA%D8%B2%D8%A9

[133] https://alrai.com/article/660247/%D9%85%D8%AD%D9%84%D9%8A%D8%A7%D8%AA/ %D8%A7%D9%84%D9%85%D9%88%D8%A7%D9%81%D9%82%D8%A9-%D8%B9%D9 %84%D9%89-%D8%A7%D8%B3%D8%AA%D9%8A%D8%B1%D8%A7%D8%AF-D8%A7% D9%84%D8%BA%D8%A7%D8%B2-%D8%A7%D9%84%D9%81%D9%84%D8%B3%D8% B7%D9%8A%D9%86%D9%8A-%D9%85%D9%86-%D8%BA%D8%B2%D8%A9

exporting its natural gas abroad. Thirdly, if the natural gas was to be exported, agreements would also have been needed to be reached with Israel, in order to allow the natural gas to pass through there, via the Israeli network/grid. Even then, the volumes mentioned in those articles would only have covered a small portion of Jordan's natural gas needs.

During that same period between 2014-2015, other baseless stories were published. Two articles, attributed to the Head of Energy Committee at the Jordanian Parliament, claimed that firstly, the Palestinian natural gas would be transported to Jordan via ships (when he would obviously know that the advanced technology required for this, is not available yet i.e. to transport natural gas by ship, without the necessary liquefaction or compression facilities!).[134] Secondly, that Jordan would buy Cypriot, instead of the Palestinian natural gas, to avoid protests in Jordan (note that thus far, no Cypriot natural gas has been produced to date in 2024, again, something he would be well aware of).[135]

Moreover, some of the Arab media exaggerate the size of the Palestinian reserves, often referring to them as 'massive' which would make Palestine energy independent.[136] This claim was made, despite the fact that these reserves are actually quite limited and, to put these numbers in perspective, if all the reserves were produced in one year, it would not be sufficient to satisfy e.g. Egypt's annual natural gas needs.

Despite the fact that in reality, the Palestinian Reserves are limited, for years the media, especially the Arab media, persists in overstating the importance of the Palestinian natural gas reserves. If no other gas reserves are discovered, the existing reserves, when developed, will be utilised for local Palestinian needs, with very

---

[134] https://www.maan-ctr.org/magazine/article/433/
[135] https://www.khaberni.com/news/%D9%82%D9%85%D9%88%D9%87-%D9%87%D8%B9
%D9%86%D9%8A%D9%88%D9%86-%D8%A8%D8%B4%D8%B1%D8%A7%D8%A1-%D8
%A7%D9%84%D8%BA%D8%A7%D8%B2-%D9%85%D9%86-%D9%82%D8%A8%D8%B1-
133128-133128
[136] https://attaqa.net/2022/07/24/%D8%AD%D9%82%D9%84-%D8%BA%D8%B2%D8%A9-
%D9%85%D8%A7%D8%B1%D9%8A%D9%86-%D9%84%D9%84%D8%BA%D8%A7%D8
%B2-%D8%A7%D8%AD%D8%AA%D9%8A%D8%A7%D8%B7%D9%8A-%D8%B6%D8%
AE%D9%85-%D9%82%D8%AF-%D9%8A%D8%B6%D9%85/

limited chances of being exported (possibly limited exports to Egypt). This means that the PNA is unlikely to play any significant role in the energy dynamics in the region.

At the time of writing, while the 2023-present Israel-Hamas war rages, there has been a huge increase in the number of articles, both printed and in the electronic media, discussing the Palestinian natural gas potential. In addition to unreliable propaganda, there have been many, war-related stories, baseless theories that there are massive reserves and that these are the reasons for the current Israel-Gaza war. Many mention, without basis, that the 'real' reason for the war is to allow Israel to control Gaza's 'massive' natural gas reserves. I am not going to delve into these stories as I have already made my position clear - that data does not support these claims, which are therefore utterly baseless. Interested readers can do a simple google search to find thousands of these stories. [137,138,139,140,141]

Furthermore, I found a huge selection of videos online, also claiming that the war is due to Israel's desire to steal both Palestinian natural gas in Gaza and oil in the West Bank. Some of the videos are slick and are professionally produced, however, they tell either pure fiction or are ignorant, showing a lack of real understanding. Some attribute all USGS potential numbers (not all are discovered yet), as natural gas reserves numbers to Gaza and then calculate a potential value of hundreds of billions of dollars to these, as yet, unfound reserves. Some go further and report unsubstantiated numbers. It is disheartening that even now, after all we have seen in recent past, the blight that is misinformation, is fuelling more hatred and this dissemination of dangerous, untrue propaganda is serving only to destabilise the region even more.

---

[137] https://theraven.substack.com/p/how-is-gaza-offshore-gas-development
[138] https://www.jordannews.jo/Section-20/Middle-East/The-true-target-is-Gaza-s-gas-the-purpose-of-Israel-s-war-34004
[139] https://www.mesp.me/2024/02/13/between-tales-and-facts-the-long-saga-of-gaza-marine/
[140] https://www.africa-confidential.com/article/id/14860/Israeli_gas_enters_Sisi%27s_Gaza_calculations
[141] https://today.lorientlejour.com/article/1357621/gas-at-stake-in-gaza-why-this-theory-doesnt-hold-true.html

Despite being a daily feature on the international news and the site of so much violence in the world, quantitatively, Palestine is a trivial player in the Eastern Mediterranean natural gas jigsaw, accounting for 0.7% of the reserves, with no production or consumption to date.

3.4 Cyprus

Cyprus is located at the heart of Eastern Mediterranean Region and experiences a lot of the problems troubling the region. The island nation has been divided since 1974, with the Greek-dominated, internationally recognised government (except by Turkey) controlling the southern two thirds of the island, while the Turkish Republic of Northern Cyprus (TRNC), recognised only by Turkey, controls the northern third. A UN buffer zone separates the two sides and years of ethnic cleansing divided the two communities totally. Many attempts at reconciling the two separate entities have failed and a solution to reunite the island remain elusive.

As a member in the EU, Cyprus enjoys the political support of the Block and has counted on this support several times against Turkey. Even though officially the whole island is part of the EU, effectively this jurisdiction applies only to the Greek controlled areas.

Two areas in southern Cyprus are controlled by the UK, as Sovereign Base Areas (SBAs) in Akrotiri and Dhekelia, they are considered to be a UK overseas territory, which remains under British jurisdiction.

Cyprus is the third largest island in the Mediterranean, but, due to the political situation on the island and the numerous parties involved, delimitating Cyprus EEZ and maritime borders is an extremely complicated issue. Figure 3.10 shows the current situation, with the numerous overlapping, conflicting claims around the island. As it stands Cyprus' EEZ claim is not recognised by Turkey and is not fully finalised by most of its neighbours. The situation is summarised as follows:

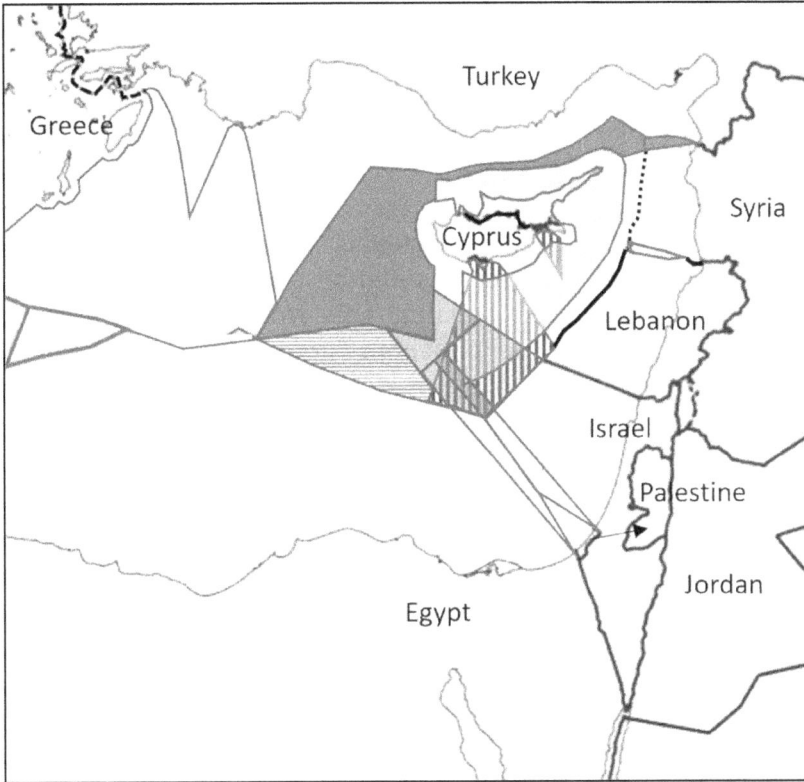

Figure 3.10: Cyrus EEZ highlighting disputed areas

Source: Drawn by author based on different sources.
Note 1: Borders are approximate and do not mean acceptance of claims.

1. The 2004 Greek Cypriot EEZ proclaimed in 2004 is not recognised by Turkey as Turkey does not recognise the Greek Cypriot government. Turkey made its position, based on 'equitable principles', that the island should not have the capacity to claim extended maritime zones when facing a bigger coastline,[142] rather it should use the continental shelf equidistant or median line principles, that Cyprus uses with other neighbouring counties. Thus, the Turkish claims overlap with the Greek Cypriot EEZ at the moment.

---

[142] https://iilss.net/maritime-boundaries-between-turkey-and-cyprus/

Note that Turkey agreed EEZ delimitation in 2011 with the Turkish Republic of Northern Cyprus (TRNC). Further claims being made by Turkey could possibly create maritime borders between Turkey, Israel and Palestine, at the expense of Cyprus.

2. If the UK decides to claim EEZ rights for its sovereign bases in Cyprus, there would be a significant overlap with the Greek Cypriot claim. This would create a potential overlap dispute between both Turkey, Northern Cyprus and the UK. However, this is politically not being considered by the British government.

3. The Israel-Cyprus EEZ is mostly covered by the bilateral agreement of 2010 based on equidistant or median line principles. However, many in Israel consider this agreement to be unfair and that Israel should have gained more areas from Cyprus. At the moment, this could only be achieved if Israel accepted Turkish territorial claims against Cyprus, which is politically improbable. Potential overlap could also exist in this scenario, if the UK decided to exercise its right to its sovereign bases EEZ.

4. If the Palestinian Authority is successful in its maritime borders and EEZ claims against Israel and Egypt, it will have maritime borders with Cyprus, but they are still indeterminate. Furthermore, if the Palestinian Authority accepted the Turkish unilateral EEZ claims and potential expansion of them, Palestine would have maritime borders with Turkey and would gain areas from Cyprus. Potential overlap could also exist if the UK were to decide to exercise its right to its sovereign bases EEZ.

5. Cyprus-Syria EEZ border is still not agreed and is covered by equidistant principles. The tripoint with Lebanon and the Southern section is yet to be determined, depending on the outcome of border delimitation between Lebanon and Syria.

6. Cyprus-Lebanon EEZ border is mostly covered by the 2007 bilateral agreement between the two countries, based on equidistant or median line principles, which is yet to be ratified by the Lebanese government. Following the agreement between Lebanon and Israel in 2022, the Lebanese and Cypriot governments have agreed to modify the tripoint so that it reflects the new maritime borders. However, depending on the outcome of border delimitation with Syria, the tripoint with Syria and the Northern section is yet to be determined.

7. Surprisingly Greece and the Greek Cypriot government have not finalised their EEZs yet.[143] Note however that, if Turkey succeed in its EEZ claims against both countries, there will be no maritime borders between Greece and Cyprus.

8. Cyprus-Egypt EEZ border is mostly covered by the 2003 bilateral agreement between Egypt and the Greek Cypriot government. The tripoint and Western section, potentially bordering Greece, are not yet determined because the border between Greece and Cyprus is not finalised. also, that area still has conflicting claims involving Turkey. Similarly, the tripoint and Eastern section, potentially bordering Israel and Palestine, are not yet determined as the border between Israel and the Palestinian Authority is not finalised. Plus, that area still has conflicting claims involving Egypt and Palestine. If Egypt accepts the Turkish unilateral EEZ claims, Egypt will have maritime borders with Turkey and will gain areas from Cyprus. In that case, potential overlap can also exist in the gained areas, if the UK decides to exercise its right to its sovereign bases EEZ.

Cyprus has minimal oil reserves, and no production. It imports all its oil needs. The country's primary energy demand has been

---

[143] https://greekcitytimes.com/2022/09/02/tragedy-of-the-cyprus-eez-unbelievable-cowardice-greece-is-afraid-to-demarcate-its-eez-with-cyprus/

relatively flat in the last two decades.

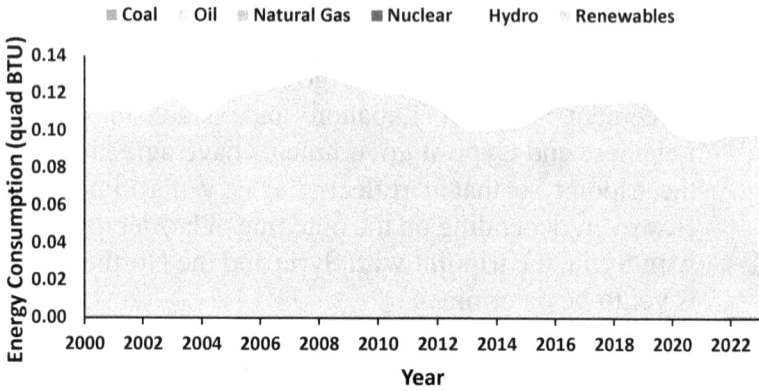

(a) Net consumption

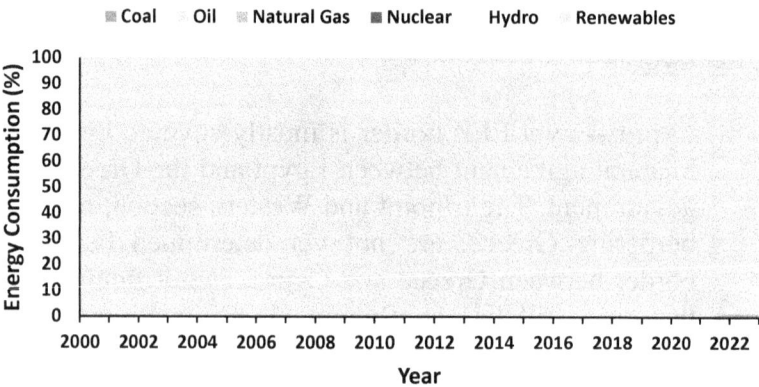

(b) Percentage consumption

Figure 3.11: Cyprus energy mix (2000 to 2023)

Source: EIA, EI

The situation regarding natural gas is different. Following years of exploration, in 2011 the first significant Cypriot field, Aphrodite, was found.[144] Several discoveries followed since including the Calypso, Cronos, Zeus and Glaucus fields.[145] However, since none of these fields are developed yet, the Cyprus energy mix still relies

---

[144] https://newmedenergy.com/operations/aphrodite/
[145] https://www.cyprusprofile.com/sectors/energy-and-environment#:~:text=These%20discoveries%20were%20particularly%20encouraging,adjacent%20to%20Cyprus'%20Block%2011.

on oil. Figure 3.11 shows the energy mix and the trends, illustrating that there is no role yet for natural gas in the energy mix yet. Although this will change soon, as the construction of a 2 mtpa LNG FSRU at Vassiliko (Vasilikos) port in Cyprus is almost completed and with the FSRU officially delivered at the yard in China in January 2024.[146] The start of the project in Cyprus continues to suffer from repeated delays,[147, 148], with a recent halt in completing the onshore construction, amidst arbitration claims by the Chinese operator.[149] The new anticipated start date is late 2024, almost 3 years delay from the original expected date.

As with reserves numbers in Israel, Cyprus' reserves numbers are underreported and currently, data sources put them at around a relatively modest 4.5 Tcf.[150, 151, 152] However, there are expectations that more discoveries will eventually materialise, as exploration momentum continues, with successful discoveries and appraisal wells likely to be announced, such as at Cronos field in 2024.[153] Note that these reserve numbers published by BP/EI, EIA and other data sources, underestimate the values significantly, since a simple summation of the major fields reserves exposes the discrepancy, giving a reserve number of 13.7 Tcf. Interestingly, the Oil & Gas Journal does not report any reserves for Cyprus. The Cypriot government believes that its EEZ holds 60 Tcf.[154]

Cyprus is not a member of GECF, but it is a founding member of the EMGF.

There are no natural gas pipelines connecting Cyprus to its neighbours at present. However, several proposed pipelines were,

---

[146] https://www.offshore-energy.biz/cosco-hands-over-first-fsru-to-cyprus/

[147] https://www.offshore-energy.biz/new-milestone-for-cyprus-lng-terminal-as-fsru-wraps-up-sea-trials/#:~:text=Floating%20storage%20and%20regasification%20unit,is%20anticipated%20in%20 July%202024.

[148] https://www.gem.wiki/Cyprus_LNG_Terminal

[149] https://cyprus-mail.com/2024/01/31/paying-for-sins-of-the-past-trouble-with-lng-terminal/

[150] http://www.cyprusprofile.com/en/sectors/energy-and-environment

[151] https://www.privacyshield.gov/ps/article?id=Cyprus-Oil-and-Gas-Exploration-and-Exploitation

[152] BGR Energiestudie 2023 - Daten und Entwicklungen der deutschen und globalen Energieversorgung, potential additional resources 17.5 Tcf.

[153] https://www.ogj.com/exploration-development/area-drilling/article/14305205/eni-confirms-extension-of-cronos-gas-discovery

[154] https://en.wikipedia.org/wiki/Energy_Triangle

or are, being considered, although only one of them may ever be realised. These are:

1. Cyprus–Egypt pipeline,[155] a planned[156] 319 km offshore pipeline to link the Cypriot Aphrodite natural gas fields to Egypt's LNG liquefaction facilities, with LNG then exported to Europe. The project was agreed in 2022,[157] but no work has started yet.

2. Cypriot section of the EastMed pipeline, linking Israel to Greece, via Cyprus and then exporting natural gas to EU, via Italy. (See Section 3.2 – for details on Israel's pipelines).

3. Cyprus-Syria pipeline – it was proposed as a link to AGP making it an export route from Cyprus to Syria and then, via Turkey, to Europe. This fantasy project never took off as, to-date Cyprus has not started a single natural gas production project to make natural gas available for export. Also, the Syrian civil war halted such potential collaborations. Plus, in the current political environment, where Turkey does not recognise Cyprus, it is unthinkable that Cyprus might allow itself to become hostage to Turkish policies by directing its natural gas through 'enemy territory'.

4. Syria-Cyprus pipeline,[158, 159] was proposed as a link to AGP to export natural gas to Cyprus via Syria, prior to the natural gas discoveries in Cyprus . The idea is cancelled as the natural gas discoveries is Cyprus rendered such a project redundant.

---

[155] Pipeline proposed capacity is 8 bcm (282 Bcf).

[156] https://enterprise.press/stories/2022/05/26/egypt-cyprus-to-break-ground-on-gas-pipeline-this-year-el-molla-72050/

[157] https://enterprise.press/ar/stories/2022/05/26/%D9%85%D8%B5%D8%B1-%D9%88%D9%82%D8%A8%D8%B1%D8%B5-%D8%AA%D8%A8%D8%AF%D8%A2%D9%86-%D8%A5%D9%86%D8%B4%D8%A7%D8%A1-%D8%AE%D8%B7-%D8%A3%D9%86%D8%A7%D8%A8%D9%8A%D8%A8-%D8%A7%D9%84%D8%BA%D8%A7%D8%B2-%D9%82/

[158] https://www.engineeringnews.co.za/article/syriacyprus-gas-link-depends-on-mideast-pipeline-2002-10-16/rep_id:4136

[159] https://www.annaharar.com/arabic/economy/oil/24082020

5. Cyprus-Turkey pipeline,[160, 161, 162] was proposed in 2019 by the TRNC government to transfer natural gas from the fields discovered offshore Cyprus in two stages, firstly to an LNG liquefaction terminal at Vassiliko (Vasilikos) in South Cyprus, then running 90 km onshore across Cyprus, then 75 km offshore, to connect to the Turkish natural gas network, where it can be sent to Europe. It is seen as an alternative to the EastMed pipeline. This project is politically highly unlikely as it would need cooperation between the two opposing factions in Cyprus, and once again, would risk Cypriot natural gas exports becoming hostage to its enemy Turkey's policies.

6. The Turkey-Cyprus pipeline was proposed as a potential route if the Blue Stream 2 pipeline took off. However, as it never progressed, beyond floating ideas, there is not much to say about it here. (See Section 3.5 – for details on Turkey's pipelines).

7. If an Israeli-Turkish pipeline were ever to happen, it would first involve significant changes in current international Israeli policies. In the unlikely event that that occurred, the pipeline would have to cross Cypriot EEZ, leading to potential conflict with the UK. It doesn't look like a realistic option at present. (See Section 3.2 – for details on Israel's pipelines).

Developing Cypriot natural gas continues to be an extremely complicated issue of conflict between the Greek Cypriot government, Turkey and with Turkey's protégé, Northern Cyprus, who are also claiming rights to this gas as well. As a result of the border disputes, Turkey is both sabotaging Cypriot plans and drill unilaterally in Cypriot waters with several stand stills occurring in

---

[160] https://www.gem.wiki/Cyprus-Turkey_Gas_Pipeline
[161] Pipeline diameter is 40" onshore and 24" offshore, proposed capacity is 3.1 bcm (109 Bcf).
[162] https://www.aa.com.tr/en/analysis/turkey-trnc-natural-gas-pipeline-to-change-geopolitical-balance-in-emed/1684563

the last few years.[163,164] These geopolitical risks can threaten the future development and progress of Cypriot natural gas resources.

At present, Cypriot development plans for its current natural gas fields have stalled. In the case of Aphrodite, the first field discovered, the Cypriot government, and Chevron and its partners continue to argue about the best plan to develop the filed. While both parties agree that they are seeking to use Egyptian infrastructure as an outlet to export the natural gas, they disagree on the actual field development plans. The Cyprus government is insisting on a floating facility in Cypriot waters, while the operators are pushing for the cheaper option of 'tie-back' to Egyptian processing facilities. At the time of writing in 2024,[165] agreement has not yet been reached. In order for this development to proceed, cooperation between all parties is key, where the benefits, such as using existing inactive infrastructure, will certainly reduce costs and be a win-win solution for everyone. Note that, in 2016, Cyprus signed an initial agreement with Egypt to transfer natural gas from its Aphrodite field to Egypt, which would then have the option to supply its domestic market or re-export the natural gas via its liquefaction plants. To date, unfortunately, no progress has been made. Also, because the field extends into Israeli water, Israeli approval is required before development can start. Israel will be entitled to partial revenue. There could be a potential link from Aphrodite field to Leviathan facilities in Israel,[166] instead of a link to Egypt, but this is considered unlikely.

Other developments are being considered for fast-tracking by ExxonMobil, Eni and TotalEnergies.[167,168]

---

[163] https://www.upstreamonline.com/politics/turkey-reignites-tensions-with-cyprus-over-bid-to-replace-russian-gas-supplies/2-1-1266181

[164] https://www.ft.com/content/69a222d4-b37c-4e7e-86dc-4f96b226416d

[165] https://www.upstreamonline.com/field-development/chevron-and-cyprus-continue-tough-talks-over-stalled-gas-project/2-1-1564991

[166] https://www.upstreamonline.com/field-development/chevron-optimises-stalled-cyprus-gas-project-with-leviathan-link-an-option/2-1-1265467

[167] https://www.upstreamonline.com/field-development/-success-exxonmobil-completes-crucial-cyprus-gas-well-to-appraise-giant-discovery/2-1-1189167

[168] https://www.ogv.energy/news-item/eni-and-totalenergies-are-swiftly-developing-natural-gas-deposits-off-cyprus-to-supply-europe

With the small Cypriot natural gas demand, the majority of the potential produced natural gas will be destined for export. Egypt remains the most logical potential market, while the EU is another potential market. At one point, it was suggested that Egypt could supply Jordan with Cypriot gas using Egyptian pipelines. However, this is no longer an option, as Egypt needs all the natural gas it can get for its own market and Jordan has secured medium-term supply deal with Israel.

The main hurdle to developing any of the Cypriot projects is finding a mutually acceptable way for the natural gas is to be exported to reach other markets. As described earlier, there have been several pipeline projects proposed, but none have been agreed thus far.

Plans for Cypriot onshore or floating LNG liquefaction export terminals are periodically considered for construction at Vassiliko, but none have got the green light so far. These export terminal projects would be a possible solution for exporting natural gas from both Cypriot and Israeli fields, with natural gas from Cronos (operated by Eni) and Glaucus (operated by ExxonMobil), as well as Israeli natural gas from Leviathan, all being prime candidates to supply such a project. [169,170]

Quantitatively, Cyprus is a trivial player in the Eastern Mediterranean natural gas game, albeit it has potential and its position could improve, if its 7% of the region's reserves are developed. However, to date there is no production or consumption. In a nutshell, with natural gas development in Cyprus, the story for the moment is 'so close and yet so far!' With so many variables and such entrenched attitudes among the Cypriots, Greeks and Turks, when it might eventually happen remains anybody's guess.

---

[169] https://www.upstreamonline.com/field-development/cyprus-pipeline-and-lng-facility-back-on-agenda-after-israeli-gas-discoveries/2-1-1435459?utm_campaign=202%E2%80%A6&utm_medium=email&utm_source=email_campaign

[170] https://www.upstreamonline.com/focus/lng-may-topple-east-med-pipeline/2-1-1375761

3.5 Turkey

As the former superpower of Eastern Mediterranean Region, Turkey still considers that it has significant entitlement in the region. It has been a presidential representative democracy, constitutional republic, with a multi-party parliamentary system for the last century but it has been edging away from democratic governance and drifting towards authoritarianism in the last few years. The current government seems to be pursuing an agenda that appears to be the quasi-restoration of the glory days of the Ottoman Empire.

Turkey is the second largest NATO member, and it is vital to the West's stability. However, it often disagrees with other Western nations, has good relationship with Russia (despite the Russian invasion of Ukraine), but has an increasingly deteriorating relationship with Israel. It has been heavily involved in the Syrian civil war.

Despite achieving EU candidate status since 1999, Turkey's accession to the block has stalled. Taking into account the animosity between Turkey, Greece and Cyprus, plus the opposition from several EU members, especially France and Austria, it is considered highly unlikely that it will succeed in achieving full member status in the EU.

Turkey has long coastlines along the Mediterranean Sea (including the Aegean Sea), in addition to the Black Sea. It also completely surrounds the Marmara Sea. Considering Turkey's EEZ claims in the Mediterranean, it is clear that Turkey, with all of its claims conflict and overlap with other claims, casts its shadow over all of the Eastern Mediterranean. Turkey, importantly, is not a signatory of UNCLOS, thus it is not bound by the convention and the agreement. In fact, Turkey has a unique, peculiar, interpretation of the concept of Continental Shelf extended borders and the equidistant or median line principles. It takes the position that, unlike the mainland, the islands have no right to EEZ. Figure 3.12 shows the disputed EEZ claims of Turkey in the region, which are as follows:

1. The Turkey-Greece dispute in the Aegean Sea which is complicated and casts permanent shadow on the bilateral relations. The two countries are at an impasse and a solution, that will satisfy both, remains elusive.

2. Outside the Aegean Sea, Turkish claims overlap significantly with the Greek claims because Turkey considers that the Kastellorizo and other Greek Islands, including Crete, have no EEZ rights.

3. Turkish claims clash with the Greek Cypriot EEZ. Turkey does not recognise the Greek Cypriot government that made its claim based on continental shelf equidistant principles and median line with other neighbouring counties. Turkey made its own claim based on 'equitable' principles, where its well-established position is, that islands should not have the capacity to claim extended maritime zones when facing a bigger coastline. [171]

   Note that Turkey agreed EEZ delimitation in 2011 with the Turkish Republic of Northern Cyprus that it does recognise.

4. There is a disputed EEZ claim between Turkey and Syria where both claims overlap. [172]

5. If Egypt were to accept Turkish unilateral EEZ claims, Egypt would have maritime borders with Turkey and would gain areas from both Greece and Cyprus.

6. Turkey and Libya agreed their bilateral maritime border delimitation in 2019. This is disputed by Greece, Cyprus

---

[171] https://iilss.net/maritime-boundaries-between-turkey-and-cyprus/

[172] Any claims to EEZ for Hatay (i.e. Iskenderun) province are excluded since this relates to a land territorial dispute which appears to be dormant, albeit not been settled, although some media outlets keep bringing it up.
See (https://syrianobserver.com/features/62207/syria-wants-iskenderun-back.html) and
(https://en.wikipedia.org/wiki/%C4%B0skenderun#cite_note-doi_10.1093/ia/iix118-18)

and Egypt,[173] who do not agree that Libya and Turkey have a joint border and consider their 2019 agreement as encroaching on their EEZ areas.

7. If Israel were to accept Turkish unilateral EEZ claims, along with their potential expansion, Israel would have maritime borders with Turkey and would gain hydrocarbon rich areas from Cyprus.

8. Similarly, if the Palestinian Authority accepts Turkish unilateral EEZ claims and their potential expansion, Palestine would have maritime borders with Turkey as well as gaining areas from Cyprus.

9. If the UK decides to claim EEZ rights for its sovereign bases in Cyprus, there is a potential overlap dispute between Turkey and the UK. There is no indication at the moment that the British government has any desire to make such a claim.

10. The Blue Homeland doctrine, advocated by some in Turkey is that Turkey should dominate the Mediterranean, reclaim the trade and maritime power once held by the Ottomans.[174, 175] Turkish President Erdoğan endorsed the policy unofficially, appearing at an event in front of the map showing the extended maritime claims in the Aegean and Mediterranean Seas.[176]

Turkey's EEZ claims in the Black Sea are more straightforward, where it shares borders with Russia, Ukraine, and Gerorgia. However, the ongoing conflict between Russia and Ukraine makes all the future borders uncertain. This is outside the scope of this book.

---

[173] https://www.greekmediagroup.com.au/gas-wars-in-the-mediterranean/
[174] https://www.euractiv.com/section/global-europe/opinion/blue-homeland-turkeys-strategy-in-the-eastern-mediterranean/
[175] https://briqjournal.com/en/the-formulation-the-blue-homeland-doctrine
[176] https://evnreport.com/politics/turkey-s-blue-homeland/

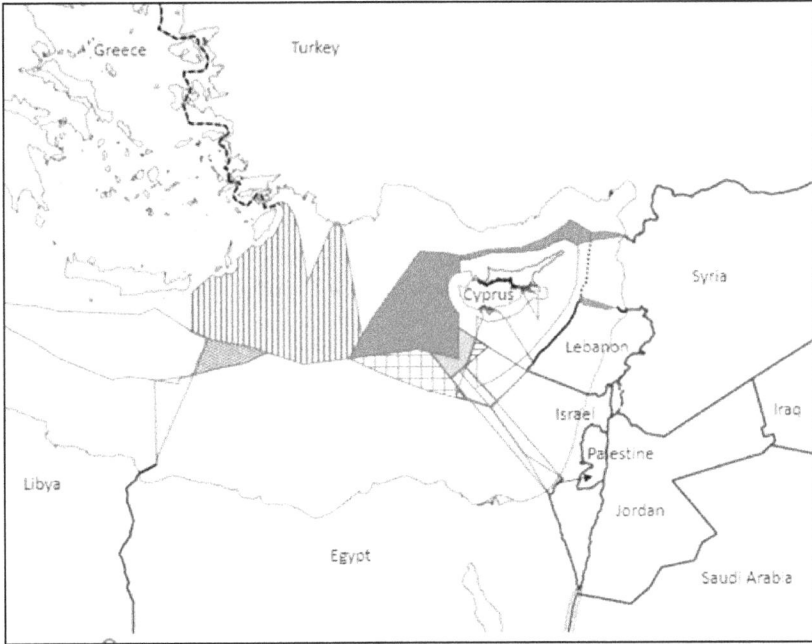

Figure 3.12: Turkey EEZ, highlighting disputed areas

Source: Drawn by author based on different sources.
Note 1: Borders are approximate and do not mean acceptance of claims.

Primary energy usage in Turkey is growing steadily and the energy sources have diversified significantly in the last two decades, when all types of energy sources grew quantitatively. Figure 3.13 shows the energy mix and the trends, illustrating the growing role of natural gas in its energy mix. However, with the country also pushing other sources of energy, including nuclear and renewables, future Turkish natural gas demand will grow in absolute terms in the medium term, although its relative importance globally, is projected to decline.

Turkey is considered to be a poor country in terms of indigenous fossil fuel energy resources, except for some deposits of low-quality lignite. Despite many years of extensive exploration, the results have yielded limited success, e.g. in 2023 it was announced that 1 billion bbl reserves were discovered in the southeastern region, close to Syrian and Iraqi borders.[177] Turkey has small oil reserves, which according to the latest estimates, its proven oil reserves stood at 507 million bbl at the end of 2023,[178] which are located onshore, mostly in the southeastern part of the country. At the end of 2023, its production averaged 80 thousand bbl/day, covering less than 8% of its needs.

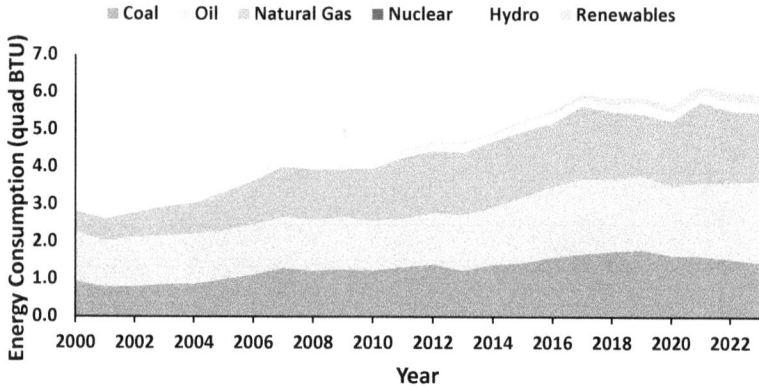

(a) Net consumption

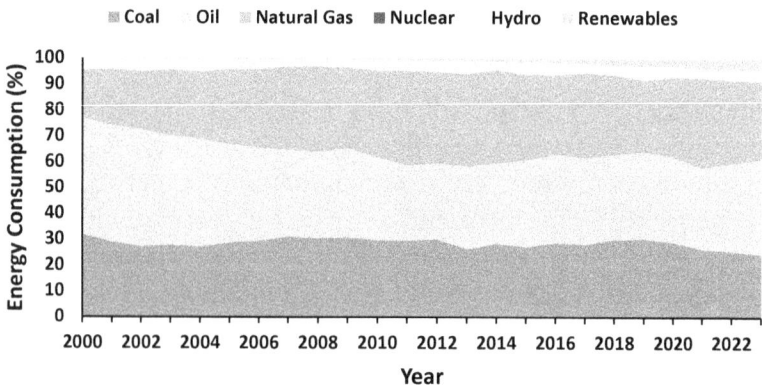

(b) Percentage consumption

Figure 3.13: Turkey energy mix (2000 to 2023)

Source: EIA, EI

---

[177] https://www.upstreamonline.com/exploration/billion-barrel-oil-discovery-marks-turkey-s-largest-ever-onshore-hydrocarbon-find/2-1-1445741

[178] https://www.ogj.com/general-interest/economics-markets/article/14302481/global-oil-and-natural-gas-reserves-both-increase

The situation was equally dire in terms of natural gas reserves, which at the end of 2023 were estimated by data sources at 109 Bcf.[179,180] In 2020 a major discovery was made in the Black Sea Sakraya field, that changed the situation (and is discussed below). After several updates, current reserves in the field are estimated at over 25 Tcf.[181] Note, the reserve numbers mentioned earlier do not include any reserves from the new discoveries. At the moment, there are blocks offered for bidding in both the Black Sea and the Mediterranean Sea, but interest is very low in the Mediterranean EEZ, partially due to the many border disputes. It is mostly Turkish companies that are performing exploration activities in internationally accepted Turkish waters, as well as in unilaterally claimed Turkish EEZ. Activities are underway in the Thrace Basin to develop unconventional natural gas resources, with potential reserves of 20 Tcf. However, the project has not produced commercial quantities yet.[182,183]

While the country is poor in oil and natural gas resources, its political leadership wants to reclaim other areas in Northern Syria and northern Iraq which are energy rich, especially Kirkuk and Mosul.

Figure 3.14 shows Turkey's annual natural gas production and consumption, demonstrating the limited production until 2023, where the spike is due to the start of production from the Sakarya natural gas field coming online. The graph also shows the rapid growth in consumption.

Turkey has a domestic cross-country natural gas pipeline network/grid connecting all major cities and other centres of population and industrial centres. It is fed from numerous import pipelines and five operational LNG regasification import

---

[179] Ibid.
[180] BGR Energiestudie 2023 - Daten und Entwicklungen der deutschen und globalen Energieversorgung, potential additional resources 58 Tcf.
[181] https://www.dailysabah.com/business/energy/turkiyes-black-sea-daily-gas-output-reaches-historic-45-mcm
[182] https://www.valeuraenergy.com/operations/turkey/thrace-basin-tight-gas-play/
[183] https://www.reuters.com/article/valeura-turkey-gas-idUKL5N1ZS1XA/

terminals, both onshore and floating. Its first terminal was the onshore Marmara Ereğlisi, which started operation in 1994 (capacity 5.9 mtpa), followed by Aliağa onshore terminal in 2006 (capacity 10.7 mtpa). The following three were FSRUs starting with Etki terminal in 2017 (capacity 5.7 mtpa), followed by Dörtyol in 2018 (capacity 7.5 mtpa) and finally Saros in 2023 (capacity 5.6 mtpa). There are plans for additional FSRU terminals, including a terminal at Ceyhan,[184] but not all are likely to materialise. Turkey has a large local distribution system to supply residential and commercial users.

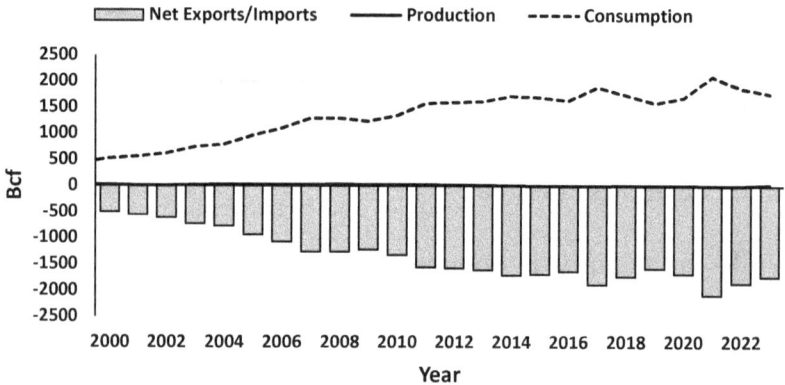

Figure 3.14: Turkey natural gas production, consumption, and trade (2000 to 2023)

Source: EIA, EI

Turkey's strategic location has made it an energy hub where numerous pipelines crisscrossed the country. It is the gateway to export Caspian natural gas and oil as far as production from Central Asia. Turkey is a minor player in production but a bigger one in logistics and transport.

Thus, capitalising on its strategic location, Turkey has deliberately positioned itself as a natural gas hub and as an indispensable gateway to the EU.[185] Its importance increased following the

---

[184] https://www.hebrewenergy.com/wp-content/uploads/2017/03/Enerji_IQ_BiWeekly_No_74.pdf
[185] https://www.upstreamonline.com/politics/erdogan-steps-up-backing-for-russia-s-turkish-gas-hub-plan/2-1-1338137?utm_campaign=2022-10-21&utm_medium=email&utm_source=email_

Russian invasion of Ukraine, as alternative natural gas sources for Europe needed to be sourced. At the same time, Russia is supporting Turkey's aim of being a natural gas hub as a way to bypass Ukraine. Currently the major cross border pipelines within Turkey are:

1. South Caucasus pipeline (SCP),[186,187,188] also known as Baku–Tbilisi–Erzurum pipeline (BTE), or Shah Deniz pipeline, is a 692 km onshore pipeline commissioned in 2006. It connects the Shah Deniz field in Azerbaijan to the Georgia-Turkey borders, where it connects to the Turkish gas network. It is considered one of the starting segments of an EU initiative, proposed in 2008, called the Southern Gas Corridor (SGC).[189] This route supplies natural gas from the Caspian and the Middle East to Europe, which reduces Europe's dependency on Russian natural gas and enables it to diversify its sources of supply. It runs parallel to the Baku–Tbilisi–Ceyhan (BTC) oil pipeline.

2. Trans-Anatolian Gas Pipeline (TANAP)[190,191,192] is the major component of SGC and is the main conduit transporting natural gas across Turkey. This onshore pipeline, commissioned in 2016, is the central part of SGC. It stretches 1841 km across Turkey, from the Georgia Border in the east to the Greek border in the west, with a branch also travelling to Bulgaria.[193] While it does not literally cross a border, it is an integral part of a cross-border route.

---

campaign&utm_term=upstrea%E2%80%A6
[186] https://www.gem.wiki/South_Caucasus_Gas_Pipeline
[187] Pipeline diameter is 42"; annual capacity was initially 8.8 bcm (310 Bcf), and is currently 24 bcm (847 Bcf), which is proposed to increase in phases to 31 bcm (1.09 Tcf) with additional compression stations. Or even to 60 billion cubic meters (2.11 Tcf) by building a second line if Central Asian natural gas is transported.
[188] https://www.bp.com/en_az/azerbaijan/home/who-we-are/operationsprojects/pipelines/scp.html
[189] https://en.wikipedia.org/wiki/Southern_Gas_Corridor
[190] https://www.gem.wiki/Trans-Anatolian_Gas_Pipeline
[191] Pipeline diameter is 56"; annual capacity is 16 bcm (565 Bcf), which is proposed to increase in phases to 40 bcm (1.41 Tcf) with additional compression stations.
[192] https://www.tanap.com/en/
[193] This was to be the stating point of the cancelled Nabucco pipeline, which was competing with TAP as a route inside the EU.

3. Trans Adriatic (TAP) pipeline, which runs from the Greek-Turkish borders to Italy, mostly onshore via Turkey, Greece and Albania, with an offshore segment crossing the Adriatic Sea. It is the third segment of SGC initiative. It was commissioned in 2020, currently transporting Azerbaijani natural gas and it has the potential to supply other Central Asian, or even Iranian, natural gas. (See Section 3.9 – for details on Greece's pipelines).

4. Turkey-Greece pipeline,[194] commissioned in 2007.[195] is a 296 km onshore pipeline connecting Karacabey in Turkey to Komotini in Greece. It was supposed to be part of a project for the transportation of natural gas from Azerbaijan to Europe via Greece and Italy. It is often referred to as the Turkey-Greece-Italy Interconnector Gas Pipeline (ITGI).

5. Tabriz–Ankara pipeline,[196,197] commissioned in 2001, [198] is a 1454 km onshore pipeline, linking the Iranian natural gas network at Tabriz with Ankara, via Erzurum, in Turkey. It is used to export Iranian natural gas to Turkey for domestic consumption. It is linked with Blue Stream pipeline a short distance before Ankara.

6. TurkStream,[199,200] or Turkish Stream, commissioned in 2018, is a strategic pipeline that links Russia with Turkey and aims to divert Russian natural gas flow from third countries (namely Ukraine). It is a 930 km pipeline that runs mostly offshore though the Black Sea and lands in Turkey west of Istanbul. It continues onshore with two branches to the Turkish-Greek border and Turkish

---

[194] Pipeline diameter is 36"; annual capacity is 11 bcm (388 Bcf).
[195] https://www.gem.wiki/Turkey-Greece-Italy_Interconnector_Gas_Pipeline_(ITGI)
[196] https://www.gem.wiki/Tabriz-Ankara_Pipeline
[197] Pipeline diameter is 40"; annual capacity is 14 bcm (494 Bcf).
[198] Reported erroneously as 2577 km in many sources on the internet.
[199] https://www.gem.wiki/TurkStream_Gas_Pipeline
[200] A double pipeline; each pipeline diameter is 40"; combined annual capacity is 31.5 bcm (1.11 Tcf).

Bulgarian border, connecting to the existing Trans-Balkan pipeline, then extending in one branch as the Balkan Stream pipeline,[201] connecting Bulgaria to Hungary.[202]

In 2021 Bulgaria, Romania, Greece, North Macedonia, Serbia, and Bosnia and Hercegovina received Russian natural gas via this branch i.e. the Balkan Stream pipeline. Note - this project, in reality, replaced the abandoned South Stream[203] project that was planned to link Russia with Bulgaria through the Black Sea and, from there, to other countries in Eastern Europe. It was cancelled in 2014 following sanctions imposed on Russia after its annexation of Crimea.

7. Blue Stream,[204,205] is a strategic pipeline commissioned in 2003, that links Russia and Turkey. It aims to divert Russian natural gas flow from third countries (namely Ukraine). It is 1213 km dual pipeline that runs, partially offshore, though the Black Sea and lands in Turkey west of Samsun, before continuing onshore to Ankara.

8. Trans-Balkan pipeline,[206,207] is a 1400 km, bidirectional onshore, pipeline linking Ukraine with Turkey,[208] with branches going to Greece and North Macedonia. It was commissioned in 1988 to deliver Russian natural gas to Romania, Bulgaria and Turkey. In 2020 the flow was reversed, where it transported Russian natural gas from TurkStream instead. The pipeline is currently also capable of receiving Azerbaijani natural gas, via TANAP and it was

---

[201] https://www.gem.wiki/Balkan_Stream_gas_pipeline
[202] https://www.upstreamonline.com/energy-security/turkey-boosts-gas-transit-capacity-for-bulgaria/2-1-1597801
[203] https://www.gem.wiki/South_Stream_Gas_Pipeline
[204] https://www.gem.wiki/Blue_Stream_Gas_Pipeline
[205] Pipeline diameter is 24" onshore, and 47" and 55" onshore; annual capacity is 16 bcm (565 Bcf).
[206] https://www.gem.wiki/Trans-Balkan_Pipeline
[207] Pipeline diameter is 42"; annual capacity is 27 bcm (953 Bcf).
[208] https://pgjonline.com/news/2023/november/ukraine-suggests-reversing-trans-balkan-gas-pipeline-flows-to-meet-demand

reported that, in 2023 it would also supply Moldova. [209],[210]

As the central transport hub in the region, several pipelines have been planned, proposed, or cancelled in the last quarter century. These include:

1. Israel-Turkey pipeline, proposed as an alternative to the EastMed pipeline, which would run offshore between Israel and Turkey. (See Section 3.2 – for details on Israel's pipelines).

2. Iraqi Kurdistan-Turkey pipeline, a proposed onshore connection, linking the planned Sulaymaniyah-Erbil-Dohuk pipeline to the Turkish natural gas network. This would enable untapped natural gas reserves in Iraqi Kurdistan, to be sold for Turkish domestic consumption or exported onwards to Europe. Despite initial negotiations between Turkey and KRG officials, significant legal hurdles exist. These obstacles are complicated by the 2022 Iraqi Supreme Court decision, stripping the KRG of independent authority over its oil and natural gas contracts, handing them back to the Federal Iraqi government. [211]

3. Iraq-Turkey pipeline, a proposed project to export Iraqi natural gas to Turkey. This would link the Turkish pipe network, then on to Europe and, in some iterations, extending further to Ceyhan on the Mediterranean. The first draft of the project was first proposed in 1967[212] and various versions of the proposals have regularly appeared since then.[213],[214] However, thus far, the project has not progressed.

---

[209] https://ecfr.eu/article/escape-routes-how-the-eu-can-help-moldova-end-its-dependence-on-russian-gas/

[210] https://www.power-technology.com/news/recean-moldova-natural-gas-reliance/?cf-view

[211] https://www.middleeasteye.net/news/turkey-iraq-kurdish-gas-easier-said-done

[212] عبد الجغيفي، صدام يوسف - سياسة العراق الخارجية في عهد الرئيس عبد الرحمن محمد عارف 17 نيسان 1966 - 17 تموز 1968: دراسة تاريخية 2021، al-Manhal lil-Nashr al-Iliktırūnī, Jordan.

[213] http://news.bbc.co.uk/1/hi/business/8054191.stm

[214] https://www.oilandgasmiddleeast.com/news/iraq-plans-to-build-gas-pipeline-to-europe

4. The interconnector between Aleppo in Syria and Kilis in Turkey is a 63 km onshore pipeline, which was completed in 2011 but has never been operational. A smaller connection already exists. This pipeline, and another alternative 15 km link can be used if rehabilitated,[215] but currently they run through rebel-held areas in Syria. This pipeline is the link between the Turkish grid and AGP.

Note that, as Iraq was considered as potential source for SGC initiative and the initial feeder to Nabucco pipeline,[216] as part of the project, a link from Iraq to AGP was agreed which was supposed to link it to the Turkish grid and subsequently to Nabucco pipeline.

It was suggested that this link could be reversed if Russian natural gas were to be exported to Syria as part of Blue Stream 2 (discussed below). This has been revived again recently, with suggestions to sell Russian natural gas to Syria and Lebanon.[217]

5. Nabucco pipeline,[218,219] was proposed in 2002. This 3893 km, mostly onshore, pipeline project from Erzurum, Turkey to Austria, via Bulgaria to several Eastern European countries would allow the EU to diversify natural gas supply and ease dependency on Russian natural gas. The pipeline's initial main supplier was to be Iraq then, in subsequent plans, supply sources shifted to Azerbaijan, potentially Turkmenistan and Egypt, via AGP.[220] Turkey suggested using Russian natural gas as part of the supply sources to the pipeline at one point, but this would contradict its objective as an substitute for Europe's reliance on Russian natural gas. The pipeline was cancelled in 2013 when Azerbaijan opted for TAP as its preferred export route.

[215] https://alkhanadeq.com/post.php?id=1318
[216] https://www.gem.wiki/Nabucco_Gas_Pipeline
[217] https://al-akhbar.com/Syria/350880
[218] https://www.gem.wiki/Nabucco_Gas_Pipeline
[219] Pipeline diameter is 56"; annual capacity is 31 bcm (1.09 Tcf).
[220] https://www.hydrocarbons-technology.com/projects/arab-gas-pipeline-agp/

6. Blue Stream 2 pipeline[221] was proposed between 2002 and 2009 as a parallel offshore pipeline to Blue Stream pipeline. The proposal suggested that two branches would be constructed, the first running to the Turkish-Bulgarian border, then on through Bulgaria to Eastern Europe, with the second running from Samsun to Ceyhan on the Mediterranean Sea. This would export natural gas to Israel (see Ceyhan-Ashkelon pipeline below), Cyprus (see Ceyhan-Cyprus pipeline below), Syria and Lebanon (via existing or proposed links between Turkey and Syria, onwards to Lebanon). The project was abandoned in favour of South Stream pipeline, running directly between Russia and Bulgaria. This was subsequently cancelled in favour of TurkStream pipeline.

7. Ceyhan-Ashkelon pipeline,[222,223] was a proposed 610 kms offshore pipeline suggested in 2009, (prior to Israel's significant natural gas discoveries), where Israel sought to import Russian natural gas to fulfil its demands. It was never a serious consideration, and the idea was cancelled once Israel made its discoveries.

8. Ceyhan-Cyprus pipeline was a possible onshore pipeline part of Blue Stream 2 proposals described above. Due to political complexities, the proposal was never seriously considered.

9. Cyprus-Turkey pipeline is proposed to link new Cypriot natural gas discoveries to Turkey via TRNC. (See Section 3.4 – for details on Cyprus' pipelines).

10. The proposed Eastring pipeline,[224,225] is an onshore pipeline running from the Turkish-Bulgarian border to

---

[221] https://jamestown.org/program/gazprom-turkey-revive-and-reconfigure-blue-stream-two/
[222] Ibid.
[223] https://www.orfonlinc.org/research/pipelines-in-the-middle-east-and-north-africa/
[224] https://www.gem.wiki/Eastring_Pipeline
[225] Pipeline diameter is TBD; annual capacity is 20-40 bcm (706 Bcf – 1.41 Tcf).

Slovakia, via Romania and Hungary. Its length would be between 1030 to 1231 km, depending on the final design. At the time of writing, the status of this project is on-hold indefinitely.

11. The proposed Persian Gas Pipeline,[226,227] also known as Pars Pipeline, Iran–Europe pipeline or Iran–Turkey-Europe (ITE) pipeline, would be a 3300 km pipeline. It would have two main sections, the first inside Iran and then internationally in Turkey and beyond. The first 1900 km part, called the Iran Gas Trunkline 9 (IGAT9),[228] was completed in 2022, but at the moment, the second section appears to have been cancelled.

12. Qatar-Turkey pipeline[229] is an unconfirmed proposal linking Qatar to Turkey to export Qatari natural gas to Europe. Two routes were suggested, the first being an onshore pipeline via Saudi Arabia, Jordan and Syria, while the second is an offshore-onshore pipeline via Kuwait and Iraq, crossing Saudi Arabian EEZ in the Persian/Arabian Gulf. It has been theorised that the former route is the real reason behind the Syrian civil war (see Section 3.8).

13. The proposed Iğdır-Nakhchivan pipeline, between Turkey and Azerbaijan would supply natural gas from TANAP in Turkey to the Nakhchivan Autonomous Republic (a landlocked exclave of the Republic of Azerbaijan), to replace Iranian supply.

In 2020, as mentioned earlier, a significant natural gas discovery was announced in the Black Sea, Sukaraya field.[230,231] There has been intensified exploration since then. The reserves of the field

---

[226] https://www.gem.wiki/Persian_Gas_Pipeline

[227] Pipeline diameter is 56"; annual capacity is 37-40 bcm (1.31 Tcf – 1.41 Tcf).

[228] https://www.gem.wiki/IGAT_9_Gas_Pipeline

[229] https://www.gem.wiki/Qatar-Turkey_Gas_Pipeline

[230] https://www.linkedin.com/pulse/sakarya-tuna-1-transformational-gas-discovery-turkey-ciaran-nolan/

[231] https://energynorthern.com/2020/09/09/turkeys-hooked-tuna-gas-discovery-could-save-it-up-to-usd-21-billion-in-import-costs/

were initially estimated at 11.3 Tcf, later increased to 19.1 Tcf and, with the discovery of additional adjacent fields, increased further to 25.1 Tcf.[232] This field development was fast tracked. It came online in mid-2023[233] and, at its peak, is projected to produce 1.4 Bcf/d to supply 30% of Turkey's annual demand.[234]

Strategically, Turkey has firmly established itself as both oil and natural gas transport, as well as an international transit hub. It is the main route for the SGC initiative to Europe sourcing natural gas from different sources outside Russia. At the same time, it is supported by Russia to be a regional hub for Russian natural gas, where it can be used to disguise natural gas that is of Russian origin and send it to Europe. While this strategy is working in the short-term, it has its long-term pitfalls, as it is the EU's long-term strategy to reduce natural gas usage (as part of its emission reduction efforts and energy transition policies).

Because of the hostile relationship with Greece and Cyprus, Turkey continues to dispute both countries EEZs and actively disrupts their exploration efforts in these areas. Meanwhile, it continues with its own efforts to explore for hydrocarbons in the same areas thus challenging the authority of both countries, creating continued tension and causing several confrontations.

Turkey is known for its stubborn policies, resilience and for using its strategic location to its advantage. Being excluded from EMGF it appears that, when it comes to energy politics in the region, it is a matter of 'Turkey against the world'. Its importance has been enhanced because of the ongoing war in Ukraine.

Quantitatively, in terms of natural gas, Turkey is a main player in the region. Turkey only accounts for 12.5% of the natural gas reserves in the region, a small (but growing) 2% of the natural gas production, but a substantial 35% of the natural gas consumption.

---

[232] https://www.aa.com.tr/en/economy/subsea-pipe-completed-to-transfer-black-sea-gas-to-shore/2865469
[233] https://www.ft.com/content/959b6f1d-8aaf-472f-9f61-7fdca4b1685e?desktop=true&segmentId=7c8f09b9-9b61-4fbb-9430-9208a9e233c8#myft:not%E2%80%A6
[234] https://www.oilfieldtechnology.com/drilling-and-production/21042023/sakarya-phase-one-kick-starts-turkeys-upstream-sector-transformation/

Even though it is a net importer, as it controls a significant volume of natural gas delivered to Europe, as well as being a large consumer itself, it is a heavy weight in terms of market dynamics.

3.6   Lebanon

Lebanon is a small nation that used to be the beacon of development and civilisation in the Arab world. Historically, the country often lived beyond its means, was plagued by and continues to endure endemic corruption in all sectors. This has created internal collapse, with the economy in free-fall and its GDP dropping more than 50% since 2019. The country's political system is in collapse and sectarian politics have made the country ungovernable. The country is a member of the Arab League and does not recognise Israel.

At the time of writing, cross-border low level fighting between Israel and Hezbollah is underway as a side tussle of the Israel-Gaza war of 2023-present.

The coast of Lebanon occupies the middle section of the Eastern Mediterranean, and its EEZ in the Eastern Mediterranean is currently governed by bilateral agreements and equidistant or median line principles. However, some of these claims are still affected by other conflicting declarations from and agreements with neighbouring countries. Figure 3.6 (in Section 3.2) shows the current situation which can be summarised as follows:

1. In 2022 Lebanon and Israel settled their long maritime border and EEZ areas dispute, after US brokered negotiations.

2. Lebanon and Syria have not settled their maritime borders yet and there is an overlap in their EEZ claims.

3. Lebanon-Cyprus EEZ is mostly covered by the 2007 bilateral agreement based on equidistant principles, but it is yet to be ratified by the Lebanese government. Following the agreement between Lebanon and Israel in 2022, the

Lebanese and Cypriot governments agreed to modify the tripoint to reflect the new maritime borders However, the tripoint with Syria and northern section are yet to be determined, depending on the outcome of border delimitation with Syria.

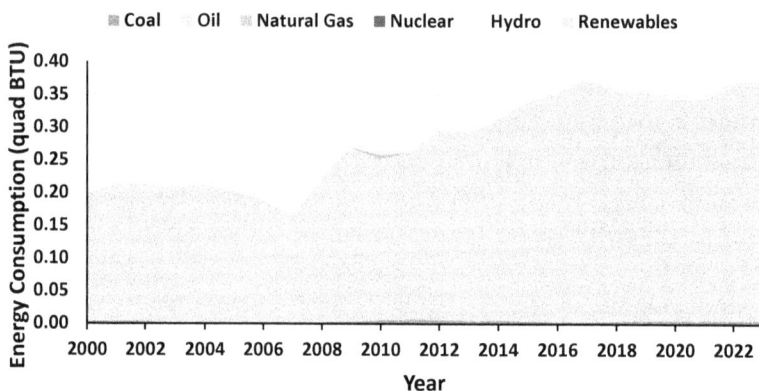

(a) Net consumption

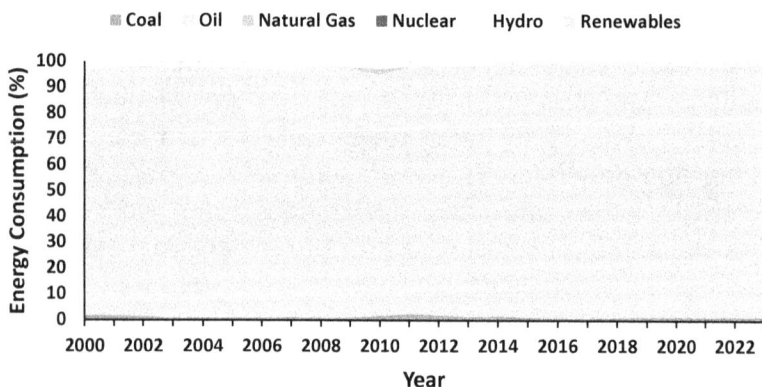

(b) Percentage consumption

Figure 3.15: Lebanon energy mix (2000 to 2023)

Source: EIA, EI

Lebanon has no oil reserves, thus no production. It imports all its oil needs. The situation with natural gas is similar, although several companies have promising expectations with their current explorations for natural gas, but with no discoveries have been confirmed yet. Lebanon's energy mix still relies on oil, and natural gas it only entered the energy market fray briefly in 2009-2010 (see Figure 3.16). Energy demand increased steadily since 2000, but after years of economic hardship, the primary energy demand started decreasing since 2017. Figure 3.15 shows the energy mix and the trends, illustrating that since 2010 natural gas had no role in the energy mix.

In 2022, an agreement was signed for Lebanon to import Egyptian natural gas, via Syria, using AGP.[235],[236] This appears to have stalled due to American sanctions on Syria. In 2022, suggestions were also for Lebanon to import Russian natural gas via Turkey and Syria, but these proposals have not been realised.[237]

Despite great optimism, to date, Lebanon has no proven natural gas reserves. The Lebanese government estimates the existing potential to be 25 Tcf.[238] Since the 2022 settlement of Israeli-Lebanese maritime dispute, active exploration is being conducted in the Qana field by a consortium led by TotalEnergies with Eni and QatarEnergy (replaced Novatek of Russia) to evaluate the volumes, although, to date the results are disappointing.[239],[240] Undeterred, the Lebanese government continues its offers, with a new bidding round launched in 2024.[241] Ultimately, any talk of Lebanon natural gas production is immature as no reserves have yet been proven there, let alone plans for developing or exporting them.[242]

---

[235] https://www.masrawy.com/news/news_economy/details/2022/6/21/2246420/%D8%B9%D8%A8%D8%B1-4-%D8%AF%D9%88%D9%84-%D9%83%D9%8A%D9%81-%D8%B3%D8%AA%D8%B5%D8%AF%D8%B1-%D9%85%D8%B5%D8%B1-%D8%A7%D9%84%D8%BA%D8%A7%D8%B2-%D8%A7%D9%84%D8%B7%D8%A8%D9%8A%D8%B9%D9%8A-%D8%A5%D9%84%D9%89-%D9%84%D8%A8%D9%86%D8%A7%D9%86-
[236] https://www.pipeline-journal.net/news/egypt-supply-natural-gas-lebanon-20-year-old-arab-pipeline
[237] https://al-akhbar.com/Syria/350880
[238] https://www.eia.gov/international/analysis/country/LBN
[239] https://www.reuters.com/world/middle-east/no-gas-finds-after-drilling-lebanons-offshore-block-9-source-2023-10-13/
[240] https://www.argusmedia.com/en/news-and-insights/latest-market-news/2540190-lebanese-exploration-blocks-in-limbo
[241] https://www.enerdata.net/publications/daily-energy-news/lebanon-launches-its-third-oil-and-gas-licensing-round-offering-9-blocks.html
[242] BGR Energiestudie 2023 - Daten und Entwicklungen der deutschen und globalen Energieversorgung, potential additional resources 21 Tcf.

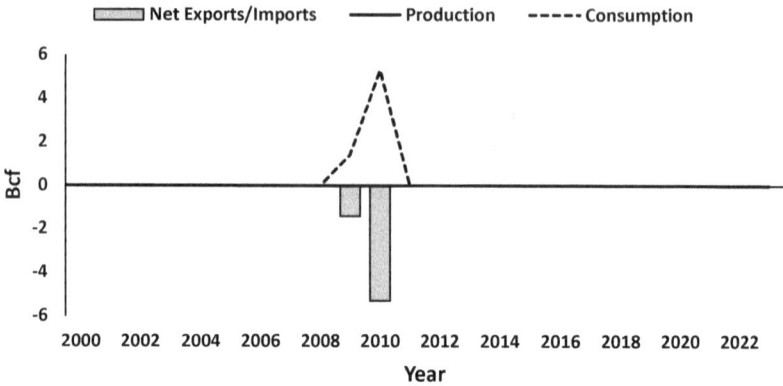

Figure 3.16: Lebanon natural gas production, consumption, and trade (2000 to 2023)

Source: EIA, EI

Currently one cross-border natural gas pipeline connects Lebanon to its neighbours. That pipeline is:

1. The Lebanese branch of AGP (see Section 3.1 – for details on Egypt's pipelines).

Other planned, proposed or cancelled pipelines include:

1. Israeli-Turkish pipeline which if it was ever built, may have to cross Syrian and Lebanese EEZs. (See Section 3.2 – for details on Israel's pipelines).

2. Syria-Sidon pipeline, there have been several cancelled[243] proposals to extend natural gas onshore pipelines to Zaharani, near Sidon in Lebanon. This proposal has included an extension from AGP[244] or from Tripoli, along the coast.

With its dire economic conditions, any natural gas discovered will be a Godsend for the country. Unfortunately, the most recent wells

---

[243] https://www.charbelnahas.org/textes/Documents_presentes_PIP/Documents_Programme_Fr_et_En/Volume_3/Project_selection_Electricity.pdf
[244] https://www.gem.wiki/Arab_Gas_Pipeline#Damascus.E2.80.93Zahrani_Extension

drilled were dry. My advice do not hold your breath – there are so many corrupt figures in charge of running the country that, even if the windfall happens, the treasury may never see a single dollar of the revenue, if and when it comes.

Note that initially, Lebanon argued that after discovery in 2010, the Leviathan field extends into Lebanese waters. However, it later acknowledged that the field lies in Israeli waters. It clearly hopes that other prospective fields in the region may be within Lebanese territorial waters. A similar claim was repeated with the Karish field, which at some point Lebanon disputed. Hezbollah even sent drones to threaten its operation in 2022.[245] The dispute over the field was resolved with a maritime border agreement in 2022.

In a nutshell, at present, quantitatively Lebanon makes no contribution to the region's natural gas reserves, production or exports, but this could change in the future.

3.7   Jordan

The small Arab kingdom is strategically located in the southern Levant. It is an oasis of stability in the region, despite the arrival of several, significantly sizeable, waves of refugees from neighbouring countries in recent years. The country's economy is weak, relying on financial support from Western and Arab countries. It has no coastline on the Mediterranean, which means that geographically, Jordan is not strictly a Mediterranean country, but as part of the Levant, it is often treated as such, politically and culturally. The country is a member of the Arab League and recognises Israel.

Jordan has very short coastline on the Red Sea where its EEZ area and maritime borders with its neighbours, (Saudi Arabia, Egypt and Israel), are stable.

Disregarding oil shale (which is in reality kerogen and is neither

---

[245] https://www.timesofisrael.com/idf-says-it-shot-down-3-hezbollah-drones-heading-for-karish-gas-field/

oil nor shale rock), Jordan has minimal oil reserves and miniscule production. It imports almost all its crude oil needs. The situation regarding natural gas is slightly different, as there is a small field is in operation. There is long history of exploration campaigns in Jordan but despite continuing governmental efforts to promote opportunities, most of which have either been disappointing or complete failures.

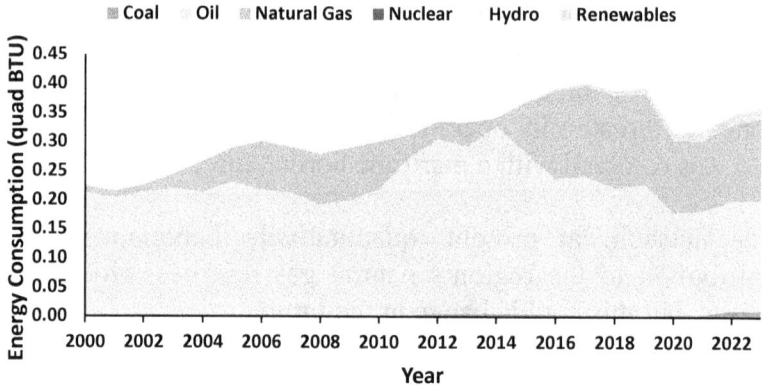

(a) Net consumption

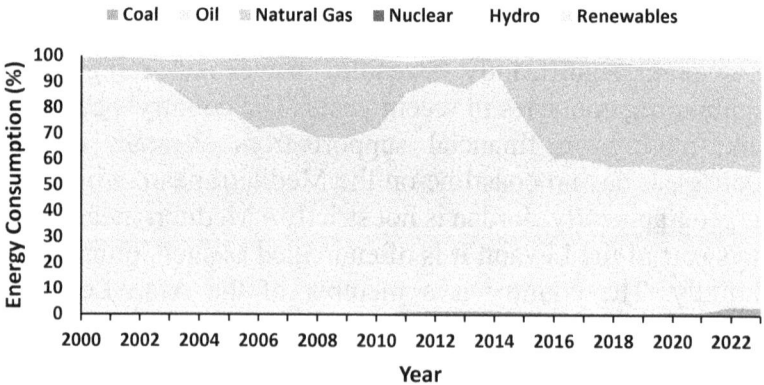

(b) Percentage consumption

Figure 3.17: Jordan energy mix (2000 to 2023)

Source: EIA, EI

Jordan has diversified its primary energy usage significantly in the last two decades. It has moved steadily from oil to rely on natural gas. After years of growth, the country's primary energy demand peaked in 2019 and declined slightly since then. Figure 3.17 shows the energy mix and the trends, illustrating the growing role of natural gas in the energy mix.

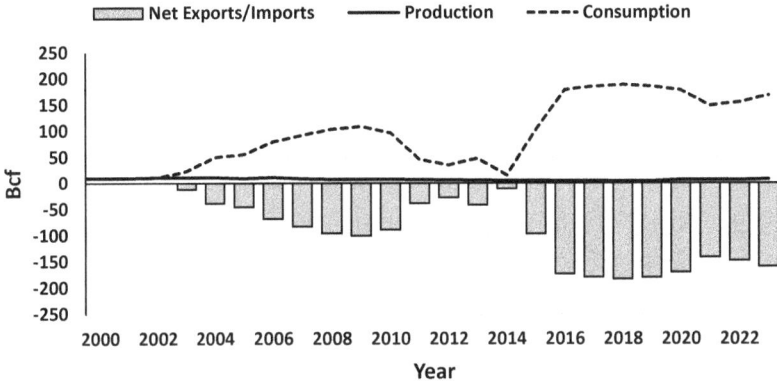

Figure 3.18: Jordan natural gas production, consumption and trade (2000 to 2023)

Source: EIA, EI

Jordan's proved natural gas reserves at the end of 2023 stood at 0.213 Tcf.[246,247] These sources are mostly located at Risha field near the Iraqi borders. This local natural gas supply started production in 1989, peaked at 30 MMscf/d in 1998, then declined gradually, as shown in Figure 3.18. By drilling new wells and adding a compression station, it underwent a rehabilitation process, increasing its production capacity to 32 MMscf/d.[248] The field's current production is just over 20 MMscf/d,[249] supplying

---

[246] https://www.ogj.com/general-interest/economics-markets/article/14302481/global-oil-and-natural-gas-reserves-both-increase

[247] BGR Energiestudie 2023 - Daten und Entwicklungen der deutschen und globalen Energieversorgung, potential additional resources 9.5 Tcf.

[248] https://www.almamlakatv.com/news/123937-%D8%A7%D9%81%D8%AA%D8%AA%D8%A7%D8%AD-%D9%85%D8%B4%D8%B1%D9%88%D8%B9-%D8%A7%D9%84%D8%BA%D8%A7%D8%B2-%D8%A7%D9%84%D9%85%D8%B6%D8%BA%D9%88%D8%B7-%D9%85%D9%86-%D8%AD%D9%82%D9%84-%D8%A7%D9%84%D8%B1%D9%8A%D8%B4%D8%A9-%D8%AE%D9%84%D8%A7%D9%84-%D8%A7%D9%84%D8%B4%D9%87%D8%B1-%D8%A7%D9%84%D9%85%D9%82%D8%A8%D9%84

[249] https://npc.com.jo/Ar/NewsDetails/%D8%A7%D9%84%D8%AE%D8%B1%D8%A7%D8%A8%D8%B4%D8%A9_%D9%82%D8%AF%D8%B1%D8%A9_%D8%AD%D9%82%D9%84_%D8%A7%D9%84%D8%B1%D9%8A%D8%B4%D8%A9_%D8%A7%D9%84%D8%A5%D9%86%D8%AA%D8%A7%D8%AC%D9%8A%D8%A9_%D8%AA%D8%B5%D9%84_%D8%A5%D9%84%D9%89_32_%D9%85%D9%84%D9%8A%D9%88%D9%86_%D9%82%D8%AF

less than 10% of the country's needs. Jordan is not a member of GECF, but it is a founding member of the EMGF.

Jordan is connected to its neighbours and has several natural gas pipelines to secure its needs. Currently these cross-border pipelines are:

1. The Jordanian section of AGP (see Section 3.1 – for details on Egypt's pipelines).

2. Israel – Jordan Interconnector (Northern Route) (see Section 3.2 – for details on Israel's pipelines).

3. Israel – Jordan Interconnector (Southern Route) (see Section 3.2 – for details on Israel's pipelines).

Other pipelines are planned, proposed or cancelled including:

1. Iraq-Jordan pipeline, running from Basra in Iraq via Haditha, to Aqaba in Jordan – cancelled. This pipeline was proposed in 2013 as an export route from Iraq to Egypt. It would have linked to AGP, or as a smaller pipeline running in parallel to the ever-delayed Iraq–Jordan oil pipeline, as a utility pipeline to power the oil pipeline. [250,251] It never progressed beyond the initial announcements and was cancelled in 2017. [252]

2. The Jordanian section of the proposed Qatar Turkey pipeline (see Section 3.5 – for details on Turkey's pipelines).

3. Jordan-Israel pipeline – linking from the Iraq-Jordan pipeline. This project is merely media speculation without feasible basis. Theoretically, once the natural gas reaches

---

%D9%853_%D9%85%D9%86_%D8%A7%D9%84%D8%BA%D8%A7%D8%B2
[250] https://www.gem.wiki/Basra-Aqaba_Oil_Pipeline
[251] https://www.rigzone.com/news/oil_gas/a/124709/iraq_begins_design_of_18b_oil_pipeline_to_jordan
[252] https://en.royanews.tv/news/11056/2017-08-07

Aqaba from Iraq, it could be sent through the reversed AGP to Israel.

As discussed earlier (see Section 3.1 and Section 3.2), Jordan has relied on importing natural gas via pipelines to supply its needs. Between 2003 – 2014 it imported gas from Egypt, then between 2018 to 2021 via AGP and, following a long-term agreement, since 2020 it imports all its needs from Israel. Please refer to the above sections for details.

When Egypt abruptly cut its natural gas supplies to Jordan in 2015, the Jordanian economy suffered a severe blow, and the government scrambled for quick solution. They fast-tracked the construction of a 3.8 mtpa LNG regasification terminal in Aqaba, the Aqaba FSRU, which operated between 2015 and 2020. However, as an insurance policy for security of supply and despite securing all its needs currently via pipelines, the Jordanian government is retaining the regasification terminal. As mentioned earlier, surplus LNG was used to supply Egypt sometimes via the reversed Aqaba-Taba link of the AGP.[253]

Despite its long-term natural gas supply contracts, the Jordanian government still seeks to develop its domestic conventional natural gas resources in the Northeast. Although the prospects of large reserves, and concessions, such as better fiscal terms, were given to several multinational companies, there has been limited success reported so far as the reserves appear to be technically challenging. Until 2014, the Risha field was the main potential for increasing Jordanian domestic production, where the government had hoped to increase domestic supply to 330 MMscf/d.[254] Unfortunately BP departed after it found no technical basis for development and its withdrawal has damaged the prospects for domestic production of natural gas.

When it was first developed, the cost of Jordanian local natural gas supply stood at less than 2$/MMBtu. However, the projected costs

---

[253] https://www.egypttoday.com/Article/3/15385/Jordan-exports-over-half-million-cubic-feet-of-LNG-to
[254] http://blogs.wsj.com/middleeast/2014/01/29/bps-risha-exit-means-slim-pickings-for-jordans-energy-needs/

for future developments were as high as over 6$/MMBtu, thus, due to these high costs, the future project was deemed unfeasible and halted by BP. Further attempts to lure foreign investment to the field did not succeed. The field is now operated by a local Jordanian operator, who in 2023, completed a rehabilitation project to increase production capacity to 32 MMScf/d. In late 2023, it was announced that there would be an increase in its capacity to 200 MMScf/d, although no details were provided. [255],[256]

Jordan has tried to establish itself as a regional natural gas hub, but efforts so far have failed. Its attempts relied on a natural gas flow, via AGP, to neighbouring countries or on natural gas exported to Egypt, from excess LNG that Jordan sourced. Instead, Israel is currently exporting natural gas to Egypt via AGP. Furthermore, rumoured natural gas pipelines connecting Qatar to the Mediterranean and, subsequently Europe via Jordan, or transporting Iranian gas, via Iraq and Jordan, to Egypt never materialised, although the latter proposal is logistically feasible, since the Iraqi and Iranian grids are already connected. With little chance of being a natural gas hub, the kingdom is now trying to become an electricity hub, connecting neighbouring countries' electrical grids, since it has overcapacity, aiming to export and facilitate power supplies.

The Jordanian media is rife with unsubstantiated and misinformed reports regarding energy markets in Jordan, especially the natural gas market. One such often published incendiary claim, is that Jordan will be paying 10-15 billion to Israel directly, for importing what is labelled as 'stolen' natural gas. This is simply untrue. The actual facts are clear and transparent. The ownership of the Israel natural gas fields is in private hands, [257] so the profits will go to

---

[255] https://www.almamlakatv.com/news/123937-%D8%A7%D9%81%D8%AA%D8%AA%D8%AA%D8%A7%D8%AD-%D9%85%D8%B4%D8%B1%D9%88%D8%B9-%D8%A7%D9%84%D8%BA%D8%A7%D8%B2-%D8%A7%D9%84%D9%85%D8%B6%D8%BA%D9%88%D8%B7-%D9%85%D9%86-%D8%AD%D9%82%D9%84-%D8%A7%D9%84%D8%B1%D9%8A%D8%B4%D8%A9-%D8%AE%D9%84%D8%A7%D9%84-%D8%A7%D9%84%D8%B4%D9%87%D8%B1-%D8%A7%D9%84%D9%85%D9%82%D8%A8%D9%84
[256] https://www.zawya.com/en/projects/oil-and-gas/jordan-to-sharply-boost-gas-field-production-rk7w9uz1
[257] The Leviathan partners (Ratio Energies 15%, Chevron Mediterranean 39.66%, and NewMed

shareholders. The cash paid goes to the operator, which is a major public company based in the USA. The revenue received by the Israeli government from the deal, comes from taxes and royalties, which will amount to around 58% of the gross revenue, minus all allowable costs, which include capital expenditure allowance, expenses, etc. This same system operates in similar deals, wherever they are made in the world. Thus, we can see that, despite what some Arab media outlets would have us believe, i.e. that this natural gas deal is a new form of 'colonialism' or control, it is, in fact, a mutually beneficial relationship between all the parties involved.

There are conspiracy theorists who believe in the existence of oil and natural gas in Jordan, where many of them claim that Jordan is awash with hydrocarbons but, for political reasons, the government is mysteriously conspiring to hide this information and keep the country poor. Unfortunately, as we have seen before with other 'misinformation' campaigns, these vocal opinions gain traction and resonate with the public, despite having no real evidence to support their arguments. Some of these so-called experts in the media, are, in reality, misleading the public.[258] They have quoted 'resources numbers' as 'reserves' (see Appendix II for clarification of the terms), or tell partial stories about shale oil reserves, etc. The stories often paint Israel as being the culprit. If the stories were true, how do they explain why Israel has not found oil itself?

Quantitatively, Jordan is a small player in the Eastern Mediterranean Region, in terms of natural gas. It is a net importer and currently only accounts for a tiny 0.1% of the natural gas reserves, a minute 0.03% of the natural gas production, and 3% of the natural gas consumption.

3.8   Syria

This is an Arab nation located on the Northeastern coast of the

---

Energy (45.34%)
[258] https://m.youtube.com/watch?v=LMThs-hyXTw

Mediterranean and is the largest country in the Levant. It has been engulfed by civil war since 2011 and the central government, led by Russian/Iranian-supported Assad, has only partial territorial control. Other territories are controlled by numerous factions, some supported by Turkey and others by the West, while others are Islamic insurgents or some are drug smuggling, criminal gangs. At the moment, a proxy war between various regional and global powers is ongoing in Syria, with regular Israeli airstrikes hitting Syrian and Iranian targets frequently, without any response from either side. The ongoing war is a major source of instability in the region and has sent millions of refugees to neighbouring countries and Europe. Syria is a member of the Arab League and does not recognise Israel.

Syria's EEZ in the Eastern Mediterranean is currently governed by equidistant or median line principles. These claims are challenged by other, conflicting, declarations and agreements from neighbouring countries. It is important to establish that Syria also is not a signatory of UNCLOS and thus it is not bound by the convention or the agreement. Figure 3.6 (in Section 3.2) shows the current situation which can be summarised as follows:

1. There is a disputed EEZ claim between Turkey and Syria, where both claims overlap.[259]

2. The Lebanon and Syria maritime borders are not yet agreed and there is an overlap in their EEZ claims.

3. Syria-Cyprus EEZ is not yet agreed and is covered by equidistant or median line principles. The tripoint with Lebanon and the southern section will be determined, depending on the outcome of border delimitation between Lebanon and Syria. Similarly, the tripoint with Turkey and the northern section is not yet determined, and will depend

---

[259] Any claims to EEZ for Hatay (i.e. Iskenderun) province are excluded since this relates to a land territorial dispute which appears to be dormant, albeit not been settled, although some media outlets keep bringing it up. See (https://syrianobserver.com/features/62207/syria-wants-iskenderun-back.html) and (https://en.wikipedia.org/wiki/%C4%B0skenderun#cite_note-doi_10.1093/ia/iix118-18)

on the outcome of border delimitation between Turkey and Syria.

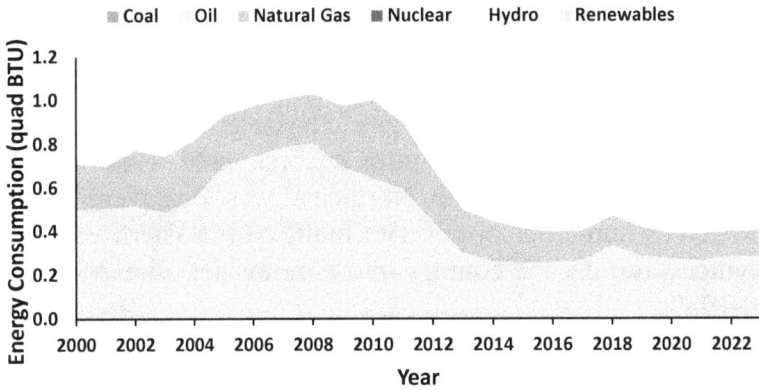

(a) Net consumption

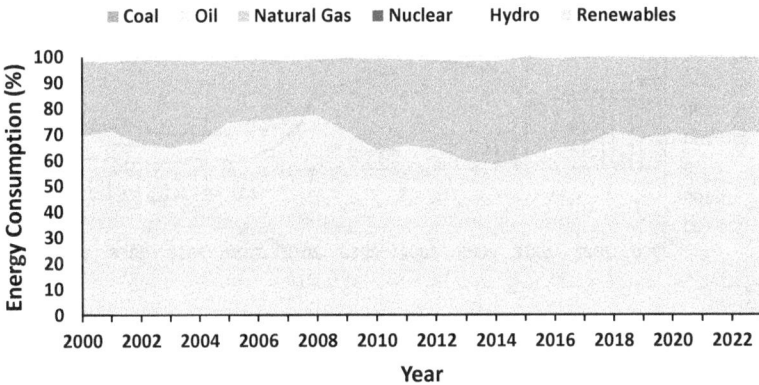

(b) Percentage consumption

Figure 3.19: Syria energy mix (2000 to 2023)

Source: EIA, EI

Syria has moderate oil and natural gas reserves and, for years, was self-sufficient supplying its energy needs. Following significant demand increase, peaking in 2008, when due to the ongoing civil war, it's energy demand collapsed after 2011, dropping by more than 50%. The current demand levels remain low and shows no signs of recovery yet. Figure 3.19 shows the energy mix and the trends, illustrating the dominance of oil in the energy mix. Natural gas plays an important role, especially in electricity generation, but quantitatively, with many power stations offline, natural gas consumption dropped significantly from its 2011 levels. The civil war has affected the size of the population, which dropped significantly, affecting oil and natural gas consumption.

Exploration for oil and natural gas in Syria goes back for decades, starting in 1933, but with little success until 1956 when the Karatchok oil field in the Northeast was discovered. It was followed by numerous discoveries in the Northeastern and Eastern provinces, turning the country into a minor net oil exporter since mid-1980s.

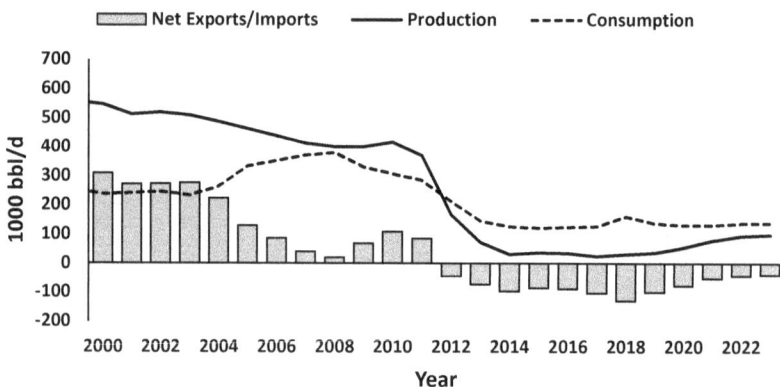

Figure 3.20: Syria oil production, consumption, and trade (2000 to 2023)
Source: EIA, EI

At the end of 2023, Syria's proven oil reserves were a modest 2.5 billion barrels.[260] Previously the country was a close third, largest producer of oil in the Eastern Mediterranean, but the civil war ravaging Syria since 2011 caused a significance drop in production. Its production in 2023 was 90 thousand bbl/day[261]. Figure 3.20 shows Syria's annual oil production and consumption, illustrating that since the Civil War started, the country has turned into net oil importer. Syria joined OAPEC in 1972, but has never

---

[260] https://www.ogj.com/general-interest/economics-markets/article/14302481/global-oil-and-natural-gas-reserves-both-increase
[261] Ibid.

joined OPEC.

Initially several international oil companies were involved in the oil industry in Syria, but most withdrew, leaving the scene dominated initially by local companies, but recently by Russian and Iranian operators.[262,263] Throughout the civil war, the producing fields changed hands several times and now many are controlled by Assad's government, some by Kurdish militias with a few controlled by the Americans but none are controlled by ISIS.

The first natural gas field in the Syria was discovered in 1968, in Soukhneh in the Palmyra Region. Several discoveries followed, mostly in the same region, with some also in the Eastern and Northeastern provinces. However, the natural gas reserves discovered were modest and all production was operated by local operators.

In terms of natural gas, Syria's proven reserves were 8.5 Tcf in 2023.[264,265] As it would be expected in current conditions, there is little exploration activity underway. New discoveries are few and far between, although one was announced in 2021, another in 2022 in the Palmyra region and further activity was announced early in 2024.[266,267]

Figure 3.21 shows Syria's annual natural gas production and consumption. This illustrates that, for most of its history, it was self-sufficient in terms of natural gas needs, but it was a net importer for few years between 2008 and 2011, when it was supplied by Egyptian natural gas via AGP.[268] In 2022, proposals were made to supply Syria with Russian natural gas, via Turkey,

[262] https://www.washingtoninstitute.org/policy-analysis/russias-energy-goals-syria
[263] https://oilprice.com/Latest-Energy-News/World-News/Iran-Considers-Forming-A-Joint-Oil-And-Gas-Company-With-Syria.html
[264] https://www.ogj.com/general-interest/economics-markets/article/14302481/global-oil-and-natural-gas-reserves-both-increase
[265] BGR Energiestudie 2023 - Daten und Entwicklungen der deutschen und globalen Energieversorgung, potential additional resources 10.5 Tcf.
[266] https://www.alahednews.com.lb/article.php?id=43101&cid=125
[267] https://en.sputniknews.africa/20231213/syria-expects-to-reclaim-us-occupied-oil-fields-soon-1064103730.html
[268] https://www.eia.gov/international/analysis/country/SYR

but no progress has been made since.[269] Unless there is a turnaround in Syria's security and economic fortunes, the prospect of importing natural gas remain non-existent.

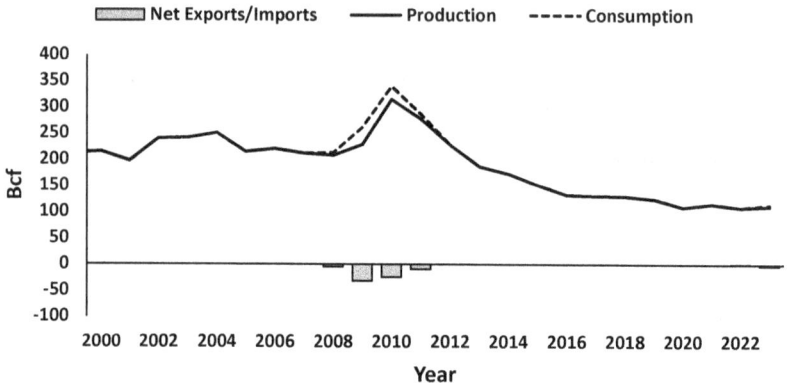

Figure 3.21: Syria natural gas production, consumption, and trade (2000 to 2023)

Source: EIA, EI

Domestically, Syria has a cross-country natural gas pipeline network/grid, connecting producing fields with main population and industrial centres. But it has no system to supply other residential and commercial users.

Despite being a net natural gas importer for only a brief period, Syria is connected to its neighbours by several natural gas pipelines, acting mostly as a transit state. Currently these cross-border pipelines are listed below:

1. The Syrian section of AGP, with connections from Jordan and to Lebanon and Turkey[270] (see Section 3.1 – for details on Egypt's pipelines). The planned pipeline stage between Homs and Aleppo was never commissioned, but the natural gas can be transported via the internal Syrian transmission network.[271]

---

[269]269 https://al-akhbar.com/Syria/350880

[270] http://www.sgc.gov.sy/ar/manager.php

[271] https://aljumhuriya.net/ar/2021/11/23/%D8%AE%D8%B7-%D8%B9%D8%B1%D8%A8%

Note that, many sections of this pipeline were sabotaged several times in the war. The flow in parts of the pipeline was reversed, as it was used to transport indigenous Syrian natural gas to demand centres.

Other planned, proposed or cancelled pipelines include:

1. Iran-Iraq-Syria pipeline,[272,273] dubbed the 'Islamic Pipeline' or 'Friendship Pipeline', is a planned onshore pipeline. It would link Iran to Syria via Iraq, it would export Iranian natural gas to Syria and Lebanon, possibly going on to Europe. The project was announced in 2010 but appears to be shelved. Iran is trying to revive the project.[274]

2. The Iraqi link to AGP,[275] was proposed in 2004. It is an onshore pipeline linking Haditha in Iraq to the Syrian natural gas network/grid and subsequently to AGP. The project has never progressed.

3. Syria-Iraq pipeline, mentioned in the media is 2021 as a possible onshore pipeline to import Egyptian natural gas to Iraq,[276] partially utilising AGP. The idea appears to have stalled as Egypt does not have the required surplus natural gas production to be exported.

4. The Syrian section of the proposed Qatar-Turkey onshore pipeline. (See Section 3.5 – for details on Turkey's pipelines).

5. Turkey-Syria interconnector, an onshore pipeline with the possibility of delivering Russian natural gas. (See Section 3.5 – for details on Turkey's pipelines).

D9%8A-%D8%B9%D8%A8%D8%B1-%D8%B3%D9%88%D8%B1%D9%8A%D8%A7/
[272] https://www.gem.wiki/Iran%E2%80%93Iraq%E2%80%93Syria_pipeline
[273] Pipeline diameter is 56", proposed capacity is 40 bcm (1.41 Tcf).
[274] https://oilprice.com/Energy/Energy-General/Why-Iran--Is-Desperate-To-Finish-This-Pipeline.html
[275] https://www.hydrocarbons-technology.com/projects/arab-gas-pipeline-agp/
[276] https://oilprice.com/Energy/Energy-General/Iraq-Revives-Its-Monster-Pipeline-Project.html

6. Cyprus-Syria pipeline, a proposed offshore pipeline to export Cypriot natural gas. (See Section 3.4 – for details on Cyprus' pipelines).

7. Syria-Cyprus pipeline, a cancelled offshore pipeline proposal to extend AGP. (See Section 3.4 – for details on Cyprus' pipelines).

8. Syria-Sidon pipeline, a cancelled onshore pipeline proposal to export natural gas to Zahrani near Sidon in Lebanon on the Lebanese coast. (See Section 3.6 – for details on Lebanon's pipelines).

9. Should an Israeli-Turkish pipeline ever be built, it may have to cross Syrian and Lebanese EEZs. (See Section 3.2 – for details on Israel's pipelines)

Since the start of the civil war, Syria was effectively outside the Eastern Mediterranean natural gas game. Natural gas activities took a back seat and, surprisingly, unlike oil production, natural gas production proved to be more resilient.

To date, all of Syria's oil and natural gas discoveries are onshore. The country has awarded exploration licences for Syria's offshore blocks to Russian companies but, to date despite this nascent involvement, no actual exploration activities have taken place. In addition, the Syrian government blames Western sanctions as a reason for the lack of progress as their sanctions prevent Syria from using advanced exploration technologies.[277]

Despite the lack of activity, there are repeated news cycles about Syria's 'hidden riches' in the Mediterranean, which make bold

---

[277] https://arabic.rt.com/business/1398830-%D9%88%D8%B2%D9%8A%D8%B1-%D8%A7%
D9%84%D9%86%D9%81%D8%B7-%D8%A7%D9%84%D8%B3%D9%88%D8%B1%D9
%8A-%D9%8A%D8%AC%D9%8A%D8%A8-%D9%85%D8%A7%D8%B0%D8%
A7-%D9%84%D8%A7-%D8%AA%D8%B3%D8%AA%D8%BA%D9%84-%D8%B3%D9%
88%D8%B1%D9%8A%D8%A7-%D9%85%D9%88%D8%A7%D8%B1%D8%AF%D8%A7%
D9%84%D8%BA%D8%A7%D8%B2-%D9%81%D9%8A-%D8%A7%D9%84%D9%85%
D8%AA%D9%88%D8%B3%D8%B7/

unfounded claims. In comparison to Jordan, conspiracy theory stories are an order of magnitude larger in Syria, being even more farfetched and often run wild. They often accuse the West and Israel, or even Russia and Iran (depending on the political affiliation of the source) of conspiring to deprive Syria of its wealth. Several fantastic stories are making the rounds in the media and include:

1. The Syrian civil war started as a result of conflict about pipelines, alleging Syria signed an agreement to export Iranian natural gas instead of Qatari natural gas. [278,279] The authors of these claims credit Syria's refusal to allow the Qatari pipeline as decisive in rescuing Russia's economy, since it allowed Russia to continue sending natural gas to Europe instead of the cheaper Qatari natural gas.

2. Syria has massive reserves, exceeding the reserves of Kuwait! [280] Or even the third largest natural gas reserves in the world, [281] and could become the third largest gas producer in the world. [282]

3. The Syrian government often under-reported oil and natural gas production, which the authors claim were far bigger, and the official numbers only revealed a fraction of true production that was declared as revenue into the budget, the rest supposedly being smuggled and sold

---

[278] https://www.almayadeen.net/Blog/%D9%85%D8%A7%D8%B0%D8%A7-%D9%83%D8%A7%D9%86-%D8%B3%D9%8A%D8%AD%D9%84-%D8%A8%D8%B1%D9%88%D8%B3%D9%8A%D8%A7-%D9%84%D9%88-%D8%B3%D9%85%D8%AD%D8%AA-%D8%B3%D9%88%D8%B1%D9%8A%D8%A7-%D8%A8%D8%B9%D8%A8%D9%88%D8%B1-%D8%AE%D8%B7-%D8%A7%D9%84%D8%BA%D8%A7%D8%B2-%D8%A7%D9%84%D9%82%D8%B7%D8%B1%D9%8A

[279] http://iraqieconomists.net/en/2016/06/06/the-secret-stupid-saudi-us-deal-on-syria-oil-gas-pipeline-war-by-f-william-engdahl/

[280] https://www.youtube.com/watch?v=0UvCUPCBsgc

[281] https://aliwaa.com.lb/%D8%A7%D9%82%D8%AA%D8%B5%D8%A7%D8%AF/%D9%86%D8%B9%D9%85%D8%A9-%D8%A3%D9%85-%D9%86%D9%82%D9%85%D8%A9-%D8%B3%D9%88%D8%B1%D9%8A%D8%A7-%D8%AB%D8%A7%D9%84%D8%AB-%D8%AF%D9%88%D9%84%D8%A9-%D8%A8%D8%A7%D8%AD%D8%AA%D9%8A-%D8%A7%D8%B7-%D8%A7%D9%84%D8%BA%D8%A7%D8%B2-%D8%A8%D8%B9%D8%AF-%D8%B1%D9%88%D8%B3%D9%8A%D8%A7-%D9%88%D8%A5%D9%8A%D8%B1%D8%A7%D9%86/

[282] https://sputnikarabic.ae/20170501/%D8%B3%D9%88%D8%B1%D9%8A%D8%A7-%D8%BA%D8%A7%D8%B2-%D8%B9%D8%A7%D9%84%D9%85-1023784061.html

secretly. Oil production numbers as big as 1.6 million bbl/d were sometimes mentioned. [283]

4. Russia took all of the block licences in order to stop Syria developing its gas, thus preventing it from competing with Russian natural gas exports. [284, 285]

With the civil war still on-going at the time of writing this book, Syria appears to be out from the race to develop hydrocarbon resources in the Eastern Mediterranean. Syria is currently the only Eastern Mediterranean country that is self-sufficient, in terms of natural gas demand. Prior to the civil war, it turned into net natural gas importer and this situation could be repeated if its economy recovers when the war ends. Thus, despite considerable natural gas reserves, it is unlikely it will become a net natural gas exporter (unless unexpectedly massive fields are discovered offshore). In the current conditions, the country's future role will be limited to a transit state that makes some money on transport fees.

Quantitatively Syria is natural gas modest player in the region with no exports or imports. It accounts for 1.3% of the reserves, 3% of the production, and a mere 2% of the consumption.

3.9   Greece

Greece was historically a dominant force in the Eastern Mediterranean, controlling the shores of most of present-day Turkey as well as other territories along the shores of the Black Sea, however, its role and territories have declined over time, reducing it now to a much less significant player. In many ways like the UK, Greece is in denial about its diminished status, continuing to view themselves as a major power, refusing to accept the new realities and its reduced role.

Despite Greece and Turkey both being neighbouring members of NATO, they are arch-enemies, and tensions between the two

---

[283] https://www.noonpost.com/content/39285
[284] https://carnegieendowment.org/sada/62037
[285] https://www.asswak-alarab.com/archives/10328

countries are a permanent feature in the region. Their longstanding acrimonious relationship is further complicated by the political situation in Cyprus (see Section 3.4) and territorial disputes in the Aegean Sea.

As a member in the EU, Greece enjoys the support of the block and has, on several occasions, enlisted this support against Turkey.

Figure 3.22: Greek EEZ, highlighting disputed areas

Source: Drawn by author based on different sources.
Note 1: Borders are approximate and do not mean acceptance of claims.

Mainland Greece has long coastlines along the Mediterranean Sea and controls most of the islands in the Aegean Sea. Greece's EEZ claims are currently governed by bilateral agreements or unilateral declarations but are widely disputed, with conflicting declarations and agreements from neighbouring countries. Figure 3.22 shows the current situation, which can be summarised as follows:

1. Greece and Italy agreed their EZZ maritime borders in

1977 and 2020.[286] The southern section and tripoint with Libya are still to be finalised. Similarly, the northern section and tripoint with Albania are yet to be finalised.

2. There is a disputed EEZ claim between Greece and Albania where both claims overlap. The disputes between the two countries may reach the ICJ.[287,288]

3. Greece's maritime borders and EEZ with Libya are governed by different interpretations of equidistant and median line principles, thus an overlap EEZ claim dispute exists.

4. Part of the Greek-Egyptian maritime borders and EEZ delimitation is governed by the 2020 bilateral agreement, giving Egypt more territories than the median line. The rest is currently governed by the equidistant or median line principles, but are disputed by Libya and Turkey. Note, if Turkey and Libya succeed in their EEZ claims there will be no maritime borders between Greece and Egypt.

5. Surprisingly Greece and Cyprus have not finalised their EEZ borders yet. Note, if Turkey succeeds in its EEZ claims, there will be no maritime borders between Greece and Cyprus.

6. The largest, most complex EEZ dispute is in the Aegean Sea between Greece and Turkey. The two countries have significant overlapping claims, where each country bases its position on different legal principles – Greece on the UNCLOS, unlike Turkey, which did not sign the agreement on the continental shelf principle. This dispute resulted in ambiguous maritime borders between Greece and Cyprus, Egypt and Libya and, despite bilateral agreements between Greece and Egypt delimiting part of

---

[286] https://sovereignlimits.com/blog/the-messy-mediterranean
[287] https://energypress.eu/war-in-ukraine-adds-to-complexity-of-greek-albanian-eez-dispute-in-ionian-sea/
[288] https://exit.al/en/greece-albania-fail-to-agree-on-maritime-borders-in-prague-meeting/

the borders, (with further unilateral border declarations by them) these borders are not recognised by Turkey or Libya. As discussed earlier, there are potentially further extreme claims by Turkey's Blue Homeland doctrine.

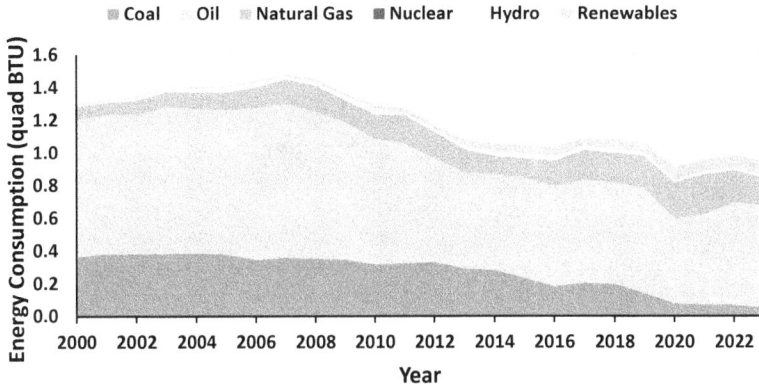

(a) Net consumption

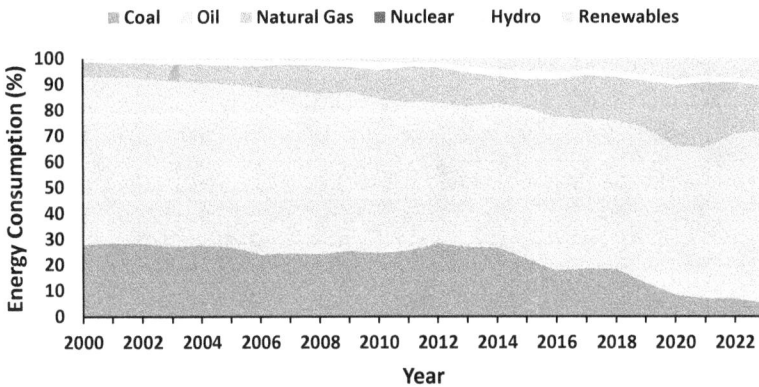

(b) Percentage consumption

Figure 3.23: Greece energy mix (2000 to 2023)

Source: EIA, EI

Greece's energy demand peaked in 2007 and has been declining since then. In the last 20 years, Greece has diversified its primary energy usage and is moving away from coal and oil. Figure 3.23 shows the energy mix and the trends, illustrating the growing role of natural gas in the energy mix.

Greece is not a traditional fossil fuel producer and is a net energy importer. Although it has a history of both oil and natural production, the volumes are small, at best supplying a tiny fraction of the country's needs. Its natural gas reserves were reported as 35 Bcf at the end of 2023.[289,290] It is a net natural gas importer getting its needs by both pipeline and LNG imports. Greece is a member of EMGF.

Figure 3.24 shows Greece's natural gas production, consumption and net trade, demonstrating a big gap in demand, which is satisfied by net imports.

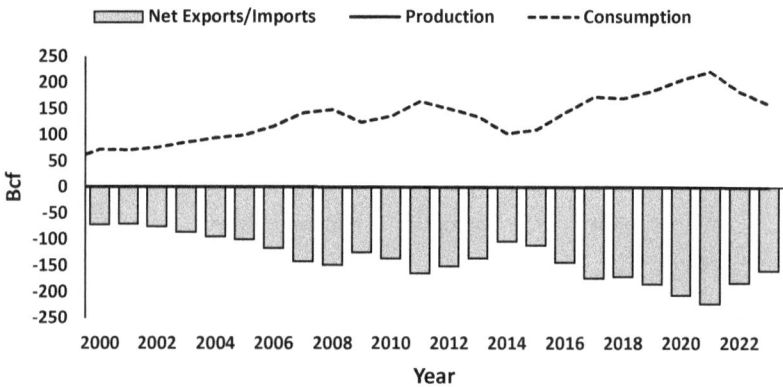

Figure 3.24: Greece natural gas production, consumption, and trade (2000 to 2023)

Source: EIA, EI

Greece has a domestic cross-country natural gas pipeline network/grid connecting all of the major cities, other centres of population and industrial centres. It is fed by numerous import pipelines and a single LNG regasification onshore import terminal at Revithoussa near Athens (capacity 4.8 mtpa). A new FSRU is starting at Alexandroupolis terminal in the northeast in 2024 (capacity 4 mtpa). Plans exist for additional onshore or FSRU

---

[289] https://www.ogj.com/general-interest/economics-markets/article/14302481/global-oil-and-natural-gas-reserves-both-increase
[290] BGR Energiestudie 2023 - Daten und Entwicklungen der deutschen und globalen Energieversorgung, potential additional resources 0.35 Tcf.

terminals, including either a terminal in Crete[291] or Thessalinki[292,293] but it is uncertain which of these may eventually will materialise. The country is also expanding its local distribution system to supply residential and commercial users.

Greece has become strategically more important to Europe as it is a transit state for current and potential major pipelines. They connect natural gas flows from the Caspian Sea and the Middle East to Eastern Europe and Italy, subsequently on to the rest of Europe. Currently, the major cross-border pipelines that exist in Greece are:

1. The Greek section of the Trans Adriatic pipeline (TAP), running onshore and extending to Northern Greece between the Turkish and Albanian borders. (See Section 3.5 – for details on Turkey's pipelines). It has been operational since 2020, transporting Azerbaijani natural gas. It will also be capable of transporting Turkmen and Iranian natural gas in the future.

2. Interconnector Turkey-Greece, is part of the cancelled Nabucco pipeline, which was part of SGC. Since 2010 it has been the operational part of the pipeline, often referred to as Greece-Italy Interconnector (IGI) - offshore, or Turkey-Greece-Italy Interconnector (ITGI) – partially onshore and partially offshore. (See Section 3.5 – for details on Turkey's pipelines).

3. Interconnector Greece-Bulgaria (IGB),[294,295] is a 182 km onshore pipeline connecting the TAP pipeline to Bulgaria, starting from Komotini, Greece. Commissioned in 2022, it is bidirectional, transporting Azerbaijan natural gas to Bulgaria and potentially beyond.

---

[291] https://www.tradewindsnews.com/news/greece-seeks-fsru-and-lng-carriers-to-serve-crete-import-project/2-1-731337?zephr_sso_ott=1lT7F5

[292] https://www.offshore-technology.com/data-insights/thessaloniki-floating-lng-regasification-terminal-greece/

[293] https://www.gem.wiki/Thessaloniki_FSRU

[294] https://igi-poseidon.com/

[295] Pipeline diameter is 32"; annual capacity is 3-5 bcm (106-176 Bcf).

4. Transit natural gas pipeline Bulgaria-Greece, is a 59 km pipeline at the Sidirokastro access point. It was commissioned in 1994 to transport Russian natural gas to Greece via Bulgaria and was reversed in 2020.[296]

Planned, proposed and cancelled pipelines are:

1. The Greek section of the proposed EastMed pipeline, (see Section 3.2 – the section on Israel's pipelines), running in the Greek EEZ from Cyprus to Crete offshore, then offshore again from Crete to mainland Greece and finally onshore, across Greece to the Ionian Sea. The offshore northwestern section (sometimes referred to as the Poseidon offshore pipeline), is technically another pipeline, connecting Greece to Italy, crossing the Ionian Sea, which replicates the cancelled IGI pipeline.

2. Poseidon onshore pipeline, the planned onshore part of the Interconnector Turkey-Greece-Italy (ITGI).[297] The future of the Greece–Italy pipeline project is unclear due to the competing of the Trans Adriatic Pipeline.[298]

3. Greece-North Macedonia Interconnector, a 56 km proposed onshore pipeline connecting TAP pipeline in Greece to North Macedonia which will transport Azerbaijani natural gas. Extending the pipeline to Serbia and Hungary is also under consideration.[299]

4. Bulgaria-Greece additional interconnectors. There are several proposals on the table including the Sofia-Sidirokastro natural gas pipeline[300] and the Kulata-Nea Mesimvria natural gas pipeline.[301] None has progressed

---

[296] Pipeline diameter is 28"; annual capacity is 3.8 bcm (133 Bcf).
[297] https://depa-int.gr/en/poseidon-pipeline/
[298] https://www.gem.wiki/Turkey-Greece-Italy_Interconnector_Gas_Pipeline_(ITGI)
[299] https://www.dailysabah.com/energy/2015/03/14/bp-becomes-a-partner-in-the-transanatolian-gas-pipeline
[300] https://www.gem.wiki/Sofia-Sidirokastro_Gas_Pipeline
[301] https://www.gem.wiki/Kulata-Nea_Mesimvria_Gas_Pipeline

beyond the planning stage so far.

5. An alternative EastMed pipeline, to run onshore along Egyptian coast, then an offshore section from Egypt to Greece, bypassing Cyprus. (see Section 3.1 – for details on Egypt's pipelines).

However, Greece's attempts to become a natural gas hub may be in vain as, the prospects of the EastMed pipeline ever actually being constructed are small and there are potentially better routes by which to export the natural gas from the Eastern Mediterranean fields in Israel or Cyprus.

Moreover, Greece's status as a transit route for natural gas from the Caspian and potentially Iran, is completely dependent on the good will of its arch-enemy Turkey. In order to build that status as a transit hub, Greece would need to keep a harmonious relationship with Turkey. Politically, Greece has pursued a smart policy when dealing with Turkey on energy issues. Whilst cooperating with Turkey in the issue of pipelines, Greece is firmly opposing Turkey's EEZ claims. This has instead turned the dispute into a wider EU-Turkey one, which has become one of the sticking points obstructing Turkey's negotiations for EU membership.

Greece formed an alliance with Cyprus and Israel to progress the controversial EastMed Pipeline. That alliance initially expanded to include Egypt, Jordan and the Palestinian Authority, with Italy and France joining later as part of the EMGF. This group is effectively an anti-Turkey alliance.

Realistically, prospects of further resources being discovered in Greek waters are limited, despite some unfounded claims by some optimists, who claim that Greece has 27 trillion m^3 (953 Tcf) natural gas reserves,[302] i.e. exceeding Qatar's reserves, without producing any evidence. Others give exaggerated, but less dramatic, estimates such as 2000 Bcm (70.6 Tcf)[303] and 3.5 Tcf.[304]

---

[302] https://slpress.gr/english-edition/an-untold-treasure-await-extraction-from-the-greek-seabed/

[303] https://greekreporter.com/2022/11/13/greece-gas-reserves-supply-eu-next-five-years/

[304] https://www.euractiv.com/section/energy/news/greece-sits-on-huge-offshore-gas-reserve-study-

Some in Greece even use the Iran-Iraq-Syria pipeline as a potential asset that would consolidate their position as a natural gas hub to Europe.[305]

In summary, Greece, Turkey and Egypt are competing to be the natural gas hub of the Eastern Mediterranean, but the Greek position is weaker than the other two as it relies on the good will of its rival and enemy, Turkey. The country will remain a net natural gas importer with little prospects of this being reversed.

Quantitatively, in terms of natural gas, Greece is a trivial player in the region. It is a net importer and only accounts for a tiny 0.02% of the natural gas reserves, a minute 0.03% of the natural gas production and 3% of the natural gas consumption.

3.10 Libya

Libya declared independence from French and British occupation in 1951, having previously been an Italian colony since 1912. In terms of area, it is the largest Eastern Mediterranean country. However, it is sparsely populated with a population of just over 7.3 million. Following the 2011 insurrection that overturned the Qaddafi regime, the country has remained in a state of upheaval. There are two governments and parliaments. Each government is supported by rival global and regional powers that are fighting a long-running proxy war in the country. Several areas in the country are governed by local militias, outside the authority of these competing governments. Libya is a member of the Arab League and does not recognise Israel.

Libya has a long coast on the Mediterranean that extends from the Eastern Mediterranean region to the Western Mediterranean. Its EEZ area is governed by bilateral agreements and equidistant or median line principles. As with all other countries in the region, its EEZ claims are challenged by other conflicting declarations and

---

says/
[305] https://greekcitytimes.com/2021/02/20/iran-iraq-syria-friendship-pipeline/

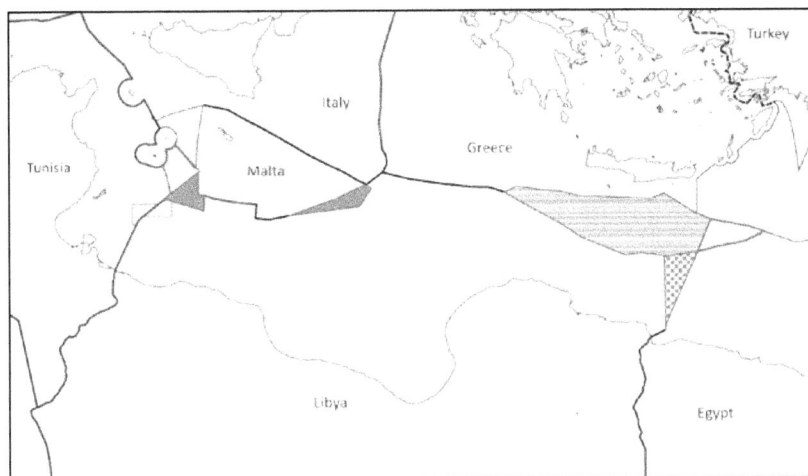

Figure 3.25: Libya EEZ highlighting disputed areas

Source: Drawn by author based on different sources.
Note 1: Borders are approximate and do not mean acceptance of claims.

agreements with neighbouring countries. Although Libya signed UNCLOS, it has not yet ratified it, thus it is not bound by its convention and agreement. Figure 3.25 shows the current situation which can be described as follows:

1. Libya-Egypt maritime border and EEZ is governed by equidistant or median line principles from Libya's perspective, but from the 2022 unilateral declaration from Egypt's perspective. Egypt's declaration is disputed by both Libya and Turkey, with an overlapping EEZ existing between Egypt and Libya.

2. Libya and Turkey agreed their bilateral maritime border delimitation in 2019. This border is disputed by Greece and Egypt, both of whom do not recognise a joint border between Libya and Turkey. They believe that the 2019 border delimitation agreement to be encroaching on their EEZ areas.

3. Libya's maritime borders and EEZ with Greece are

governed by different interpretations of equidistant or median line principles. As a result, an overlap EEZ dispute exists.

4. Libya's maritime borders and EEZ with Italy are governed by equidistant or median line principles, but the borders are not formalised.

5. The maritime borders between Libya and Malta are partially determined by a ruling from ICJ in 1985. The remaining borders are by different interpretations of equidistant or median line principles, as a result, two overlap EEZ disputes exist. Furthermore, part of the border is affected by an overlap EEZ dispute between Malta and Tunisia.

6. Libya and Tunisia agreed their bilateral maritime border delimitation in 1988 following an ICJ ruling in 1982. The bilateral agreement also established a joint development area between the two countries. A section of this border is affected by two EEZ disputes between Malta, Libya and Tunisia.

Libya possesses significant oil and natural gas reserves. Its energy demand is dominated by hydrocarbons which, because of huge subsidies, are provided cheaply. As a result, no other energy sources have made any breakthrough, despite the potential for renewables, especially solar energy. The turmoil engulfing Libya since 2011 has created economic instability, which has resulted in its energy demand declining. However, despite this, the share of natural gas has increased steadily in the last two decades at the expense of oil. Figure 3.26 shows the energy mix and the trends, illustrating the growing role of natural gas in the energy mix.

Fossil fuel extraction, the development of oil and the natural gas sector in Libya has a rich long history that goes back several decades before any natural gas discovery in the Eastern Mediterranean basins. Exploration started in 1956 and, following minor discoveries in 1957 in Ghat province, the first significant oil discovery was recorded in 1959 onshore at Zelten, followed in 1971 by a discovery offshore in Western Tripolitania.

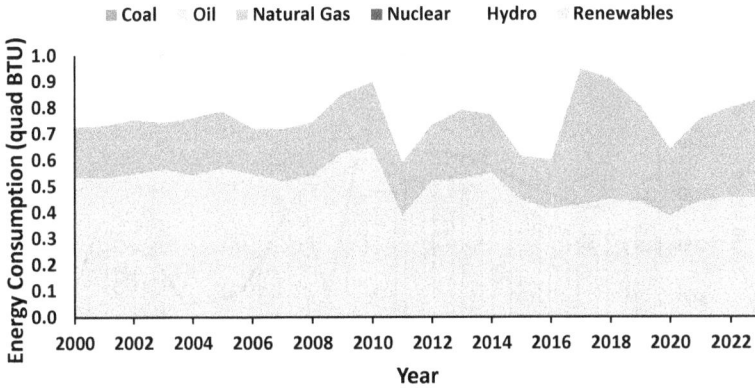

(a) Net consumption

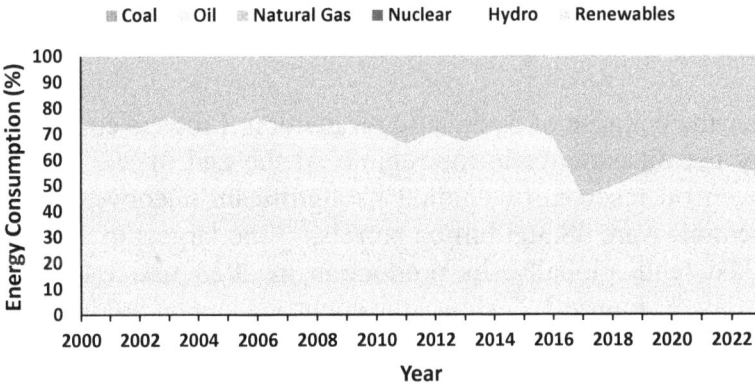

(b) Percentage consumption

Figure 3.26: Libya energy mix (2000 to 2023)

Source: EIA, EI

While most of Libya's producing basins are not strictly part of the Eastern Mediterranean designation, the direct involvement of Egypt, Greece and Turkey during the second Libyan civil war have dragged Libya firmly into the Eastern Mediterranean 'playing field'.

Although this involvement does not affect current discoveries in the Eastern Mediterranean, the aftermath resulting from this could have implications on future pipelines, drilling and exploration plans, as well as potential EEZ disputes in the future.

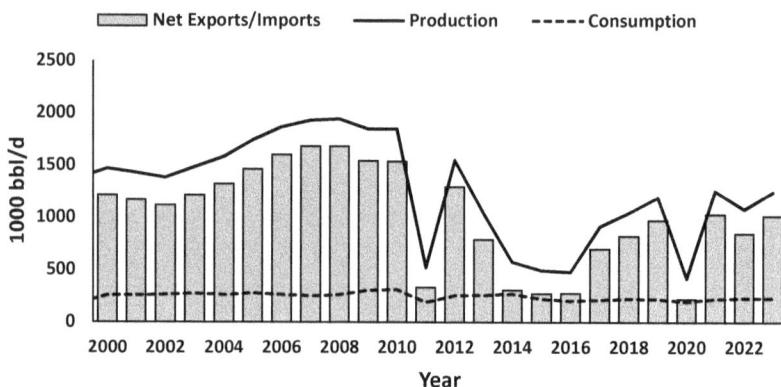

Figure 3.27: Libya oil production, consumption, and trade (2000 to 2023)
Source: EIA, EI

After the collapse of Syria's oil production, Libya is currently the only net oil exporter in the region. At the end of 2023, Libya's proven oil reserves (excluding its significant unconventional oil potential) were 48.363 billion barrels,[306] the largest in Africa and ranked tenth globally. Its production in 2023 was 1.15 million bbl/day.[307] Figure 3.27 shows the country's annual production and consumption, illustrating that that the country remained a net exporter, despite significant periods of collapse in oil production in a situation that remains volatile following the 2011 civil war. Libya was a founding member of OAPEC in 1968. It joined OPEC in 1962 and retains both memberships. Note, due to the on-going civil war, Libya is exempted from OPEC+ production quotas.

In terms of natural gas, the first discovery onshore was recorded in 1959 in the Ghadames Basin. Offshore, the first natural gas was discovered in the Gulf of Sirte in 1967.

---

[306] https://www.ogj.com/general-interest/economics-markets/article/14302481/global-oil-and-natural-gas-reserves-both-increase
[307] Ibid.

Libya's proven natural gas reserves (excluding its significant unconventional potential) were 53.144 Tcf at the end of 2023,[308,309] the fifth largest in Africa and ranked 22nd globally. As with Egypt's position, its ranking in Africa is dropping following recent discoveries in Mozambique and other African countries.

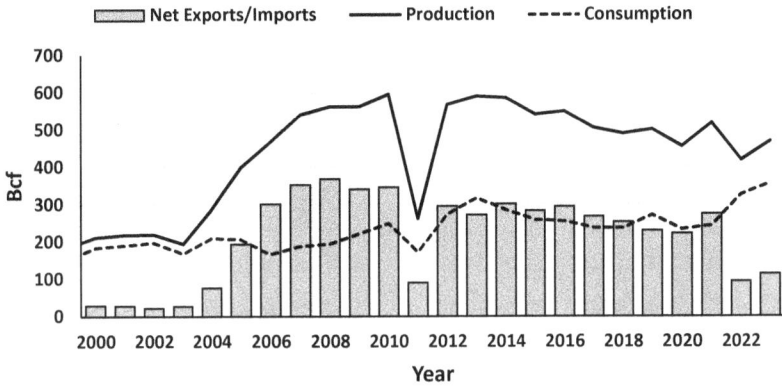

Figure 3.28: Libya natural gas production, consumption, and trade (2000 to 2023)

Source: EIA, EI

Figure 3.28 shows Libya's annual natural gas production and consumption, demonstrating resilient natural gas production and net exports, unlike the oil situation. Note, a significant proportion of this production is offshore, hence it suffered less disruption from the upheaval affecting onshore fields. Libya was a founding member of GECF.

At the moment, Libya's natural gas production is used domestically and exported to Italy via an offshore pipeline. In 1970 Libya was one of the earliest countries to establish a 3.2 mtpa LNG liquefaction facility, at Marsa El Brega, after only Algeria[310] and the USA.[311] However, its LNG terminal was mouthballed (i.e.

---

[308] https://www.ogj.com/general-interest/economics-markets/article/14302481/global-oil-and-natural-gas-reserves-both-increase
[309] BGR Energiestudie 2023 - Daten und Entwicklungen der deutschen und globalen Energieversorgung, potential additional resources 232 Tcf.
[310] Azrew-Skikda LNG
[311] Kenai LNG in Alaska

effectively shut down) in 2011 war and has not been restarted yet.

The majority of oil and natural gas production in Libya is controlled by the Libyan National Oil Corporation. However, some resources are produced as joint ventures or exploration and production sharing agreements (EPSAs) with IOCs, such as Eni, OMV and TotalEnergies, as well as other NOCs, such as Sonatrach.[312]

Domestically, Libya has a cross-country natural gas pipeline network/grid connecting the producing fields with its main population and industrial centres. There is also a major coastal trunkline connecting Benghazi and Tripoli. Libya has no distribution system to supply residential and commercial users at present.

Currently several cross-border natural gas pipelines connect Libya to its neighbours. These pipelines are:

1. Greenstream pipeline,[313,314] a 560 km offshore pipeline connecting Mellitah in Libya and Gela in Sicily, Italy crossing the Mediterranean Sea, and was commissioned in 2004. It is fed by Bahr Essalam offshore field, and onshore Bouri and Wafa fields near the Algerian border. The latter fields are connected via internal/domestic 530 km onshore pipeline to Mellitah. A proposed expansion to the line was ruled out in 2022.[315]

Other planned, proposed, or cancelled pipelines include:

1. Egypt-Libya pipeline, was a cancelled pipeline to link

---

[312] https://noc.ly/index.php/en/companies-2
[313] https://www.gem.wiki/Greenstream_Pipeline
[314] Pipeline diameter is 32"; annual capacity is 11 bcm (388 Bcf).
[315] https://www.libya-al-mostakbal.org/88/43423/%D9%86%D9%88%D9%81%D8%A7-%D9%85%D9%86-%D8%BA%D9%8A%D8%B1-%D8%A7%D9%84%D9%85%D8%AD%D8%AA%D9%85%D9%84-%D8%A8%D9%86%D8%A7%D8%A1-%D8%AE%D8%B7-%D8%A3%D9%86%D8%A8%D9%88%D8%A8-%D8%BA%D8%A7%D8%B2-%D8%AC%D8%AF%D9%8A%D8%AF-%D8%A8%D9%8A%D9%86-%D9%84%D9%8A%D8%A8%D9%8A%D8%A7-%D9%88%D8%A5%D9%8A%D8%B7%D8%A7%D9%84%D9%8A%D8%A7.html

Egyptian natural gas network to Greenstream pipeline, to utilise the Libyan internal transmission network. (See Section 3.1 – for details on Egypt's pipelines).

2. An alternative EastMed pipeline was proposed in 2022[316] to construct an onshore pipeline from Egypt to Libya, then to an offshore segment, to Greece. The route via Egypt and Greece is, in engineering terms, less technically challenging, as it avoids major geologic faults and runs in shallower water.[317] However, in 2022 the Libyan government based in Tripoli called a pipeline between Libya and Greece useless, which does not bode well for its future.[318]

3. Nigeria-Libya pipeline, an alternative route to transport Nigeria's natural gas,[319] was floated in 2022 by the Libyan Oil Ministry, to counter two other suggested routes i.e. via Algeria or via the West African Coast and Morocco.

4. Mellitah-Gábes pipeline,[320,321] was a 286 km onshore pipeline linking Libya and Tunisia, which was initially proposed in 1997,[322] progressed to engineering in 2006, but has since then abandoned.[323]

5. Libya-Malta pipeline,[324,325] was proposed in 2003 as a connection to Greenstream pipeline, but an agreement was never reached.[326] The project was revived in 2021 with a

---

[316] https://libyaalahrar.net/noc-chief-says-libya-studying-gas-pipeline-project-with-greece-and-egypt/

[317] https://www.agenzianova.com/en/news/gasdotto-egitto-libia-grecia-nuova-rotta/

[318] https://libyaupdate.com/dbeibehs-government-creating-libyan-gas-pipeline-with-greece-is-useless/

[319] https://www.libyaherald.com/2022/09/libyas-oil-ministry-counter-proposes-that-nigeria-europe-gas-pipeline-run-through-libya/

[320] https://www.gem.wiki/Mellitah-G%C3%A1bes_Pipeline

[321] Pipeline diameter is 24"; annual capacity is 2 bcm (70 Bcf).

[322] https://www.projectfinance.law/publications/2004/june/libya-poised-to-be-a-major-gas-exporter/

[323] https://www.meed.com/libya-tunisia-gas-pipeline-impediments-to-progress/

[324] https://euobserver.com/rule-of-law/153765

[325] https://timesofmalta.com/articles/view/malta-to-be-linked-to-libya-italy-gas-pipeline.149406

[326] https://www.maltatoday.com.mt/comment/blogs/1178/when-greed-killed-a-gas-pipeline#.Y9ugcnbLeUk

proposed 159 km link from Gela, in Sicily, to Malta.[327]

6. Libya-Egypt pipeline, linking with Egyptian LNG facilities was proposed in 2022, again in 2023, but no firm plans were put together, so no progress has been made.[328] There is no prospect of the project advancing unless the political situation in Libya stabilises and conditions become secure once again.

Despite Cyrenaica being geographically located in the East of Libya, making it an integral part of the Eastern Mediterranean region, Libya's current natural gas activities are in the distant West of the country and have not been connected to the Eastern Mediterranean network. However, Libya's position is changing rapidly, especially on the political arena, bringing it firmly into the Eastern Mediterranean natural gas game as a potentially major player. The controversial maritime border agreement with Turkey is one example of the importance of Libya's role and the rival governments involved, now recognize different versions of maritime borders. Unfortunately, the 2022/2023 proposals for alternative natural gas pipelines to Europe, from the Eastern Mediterranean, via Libya, have become another point of contention between the rival Libyan governments, and this continues to be the case.

Despite its political involvement in Eastern Mediterranean affairs and the potentially huge benefits of connecting the Libyan natural gas network with other Eastern Mediterranean countries, no tangible work has been done to date. As things stand, Libya remains detached from the rest of the region in natural gas market terms.

Quantitatively, even though it is a net natural gas exporter, the country is effectively a benched player in the region. The country accounts for a substantial 27% of the natural gas reserves, a

[327] https://www.businesstoday.com.mt/people/people/1668/paradise_regained__a_gas_pipeline_to_italy#.Y9uiPnbLeUk
[328] https://worldview.stratfor.com/article/external-meddling-libyas-crisis-portends-more-infighting-and-oil-blockades

significant 15% of the natural gas production, and 7% of the natural gas consumption.

Chapter 4
THE SUPPORTING PLAYERS AND THEIR ROLES

In the previous chapter I looked at the East Mediterranean region's natural gas situation, portraying it as a game where I analysed the main 'players' in the game. In this chapter, let's extend the analogy to include the other supporting 'players' influencing events in the region, i.e. the natural gas resource management, policies and future.

Collectively, these peripheral countries are bigger than 'the East Med Ten'. Although there are many countries that could be considered as relevant, for the purpose of this book, only the ones with major influence are included. Despite their inclusion in 'the game' they are mostly unaffected by what happens in the Eastern Mediterranean.

In this chapter I could provide in-depth analysis into the details of the natural gas related issues in these countries e.g. pipelines details or EEZ disputes but that is perhaps more suitable for another book. Instead, I will provide a thumbnail sketch of issues relevant to the Eastern Mediterranean region.

4.1   Italy

During the Roman Empire times, the Mediterranean Sea, was in its entirety, a vast Roman lake, where the Rome controlled all its shores, including all of what is now the Eastern Mediterranean region. Although Italy has shrunk considerably, it still has significant ties to the Eastern Mediterranean Region, politically, economically and culturally. Italy is a founding member of both NATO and the EU.

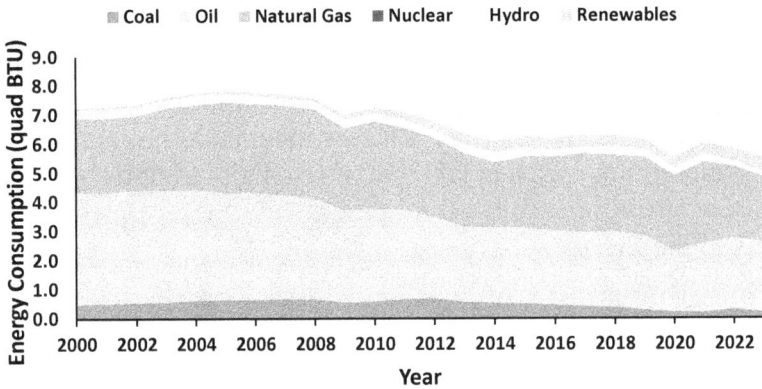

(a) Net consumption

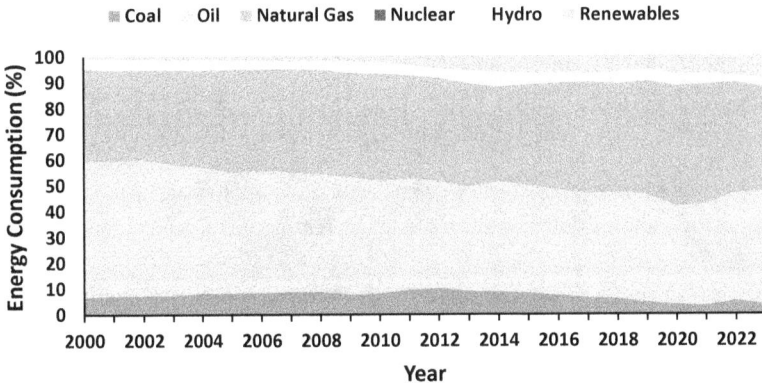

(b) Percentage consumption

Figure 4.1: Italy energy mix (2000 to 2023)

Source: EIA, EI

Italy is the fourth largest European energy consumer behind Germany, France and the UK. However, in terms of indigenous oil and natural gas reserves, the country is poor, with an estimated volume standing at 538.682 million bbl and 1.315 Tcf respectively.[1,2] It imports most of its needs. The Italian primary energy mix is shown in Figure 4.1, which shows that natural gas

---

[1] https://www.ogj.com/general-interest/economics-markets/article/14302481/global-oil-and-natural-gas-reserves-both-increase

[2] BGR Energiestudie 2023 - Daten und Entwicklungen der deutschen und globalen Energieversorgung, potential additional resources 14.2 Tcf.

consumption and share is growing.

Domestic Italian natural gas production stood at 99 Bcf in 2023, accounting for only 4.5% of the total consumption (see Figure 4.2). Clearly, this means that Italy is a pure natural gas importer. Due to its consumption, size and location, it is a major importer as well as a gateway for imported natural gas, mostly by pipelines, to other parts of the EU.

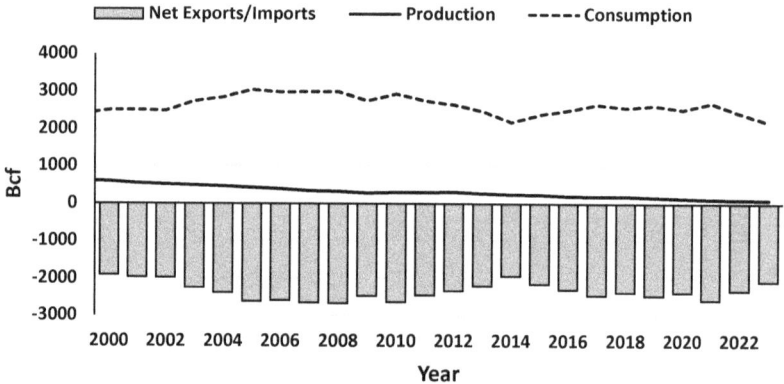

Figure 4.2: Italy natural gas production, consumption and trade (2000 to 2023)

Source: EIA, EI

This makes Italy a natural market for Eastern Mediterranean natural gas as a geographically close major consumer. The Italian international oil and gas company, Eni, which owns and operates natural gas assets in the Eastern Mediterranean is amongst the main players.

In the last few years, the Italian government has participated in meetings regarding potential Eastern Mediterranean pipelines, and even joined the EMGF. It expressed interest in the stalled EastMed pipeline between Israel and Greece via Cyprus and its potential link to Italy. But despite political statements of support, no tangible work has been forthcoming from the Italians to push the project further. In 2023 Eni commented that the EastMed Pipeline project will not proceed without Turkey, which angered its promoters.[3]

Italy is expanding its LNG regasification import capacity. There are currently 3 onshore and 3 offshore terminals, with a total capacity of 17.2 mtpa. There are other projects, totalling 5 mtpa, under construction and 26.9 mtpa are proposed or stalled, although it is unlikely that all of them would proceed. Italy historically imported LNG from Egypt and, after an interruption of several years, resumed this in 2022. Thus, if the natural gas from the Eastern Mediterranean becomes available and is sold as LNG, leading to more liquefaction terminals being constructed, Italy could be an interested buyer as the transport costs and delivery times would be very competitive.

Italy is positioning itself as a natural gas hub and gateway to the EU. Following the Russian invasion of Ukraine, its importance has increased as a gateway for alternative natural gas resources to Europe to substitute for the sanctioned Russian natural gas.

Italy has an extensive internal domestic natural gas pipeline network/grid. It is also linked to major cross border pipelines with most of its neighbours:

1. Trans-Mediterranean pipeline from Algeria, via Tunisia, to Sicily.
2. Greenstream pipeline from Libya, via Maltese EEZ, to Sicily. (See Section 3.10 – for details on Libya's pipelines).
3. Trans Adriatic (TAP) pipeline via Turkey, Greece and Albania. It currently transports Azerbaijani natural gas and has the potential to supply other Central Asian, or even Iranian, natural gas. See Section 3.5 – for details on Turkey's pipelines).
4. Trans Austria Gas (TAG) pipeline, a bi-directional pipeline linking Austria to Italy.[4]
5. Italy-Slovenia interconnector.[5]

---

[3] https://cyprus-mail.com/2023/05/19/no-eastmed-pipeline-without-turkey-eni-says/
[4] https://www.gem.wiki/Trans_Austria_Gas_Pipeline
[5] https://www.gem.wiki/Rogatec%E2%80%93Gorizia_Gas_Pipeline

6. Trans Europa Naturgas Pipeline, a bi-directional pipeline linking Germany to Italy, via Switzerland.[6]

Potential, proposed and cancelled pipelines include:

1. EastMed pipeline. (See Section 3.2 – for details on Israel's pipelines).
2. Hungary-Slovenia-Italy Interconnector gas pipeline.[7]
3. Italy-Croatia interconnector.[8]
4. Gasdotto Algeria–Sardegna Italia (GALSI) pipeline, connecting Algeria to Italy directly via Sardinia.[9] This pipeline has been shelved.

Despite Italy's ambition to be a European natural gas hub and gateway, there are no interconnections to France. France, due to its nationalistic internal policies, has prevented the connection of its grid with Italy and Spain. However, being responsive to the change in the global geopolitical landscape, the Italian government is concentrating on developing energy ties with Africa and encouraging Eni to make more investments there.[10]

4.2   France

France is a major European global power, with nuclear weapons and veto power at the UN Security Council. Despite its relatively small geographical size, its colonial history as a significant empire means that its political and military influence often extends far beyond its current borders. France is a founding member of both NATO and the EU.

Despite being the second largest European primary energy consumer,[11] since it relies on nuclear energy for the majority of its

---

[6] https://www.gem.wiki/Trans_Europa_Naturgas_Pipeline
[7] https://www.gem.wiki/Hungary-Slovenia-Italy_Interconnector_Gas_Pipeline
[8] https://knowledge.energyinst.org/search/record?id=110109
[9] https://www.gem.wiki/GALSI_Pipeline
[10] https://www.ft.com/content/05d17d35-b0c3-47d2-b6b7-
6f7d65d758fc?desktop=true&segmentId
=dd5c99e9-30be-ddd0-c634-ff3a0c2b738f#myft:n%E2%80%A6
[11]   https://www.statista.com/statistics/332520/primary-energy-consumption-in-selected-countries-in-europe/

electricity needs, France consumes less natural gas than other major European countries (i.e. Germany, the UK and Italy). Figure 4.3 shows France's energy mix.

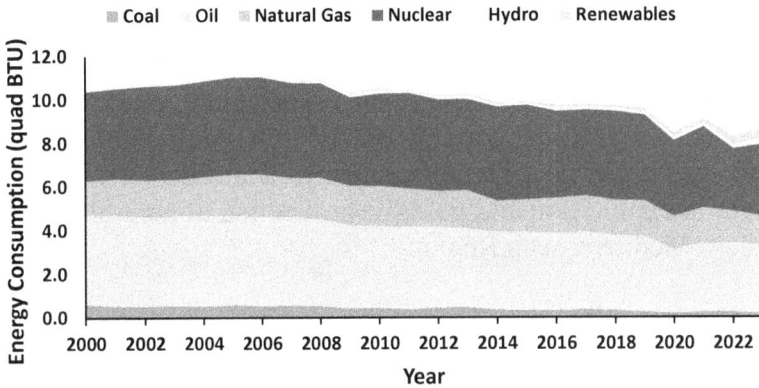

(a) Net consumption

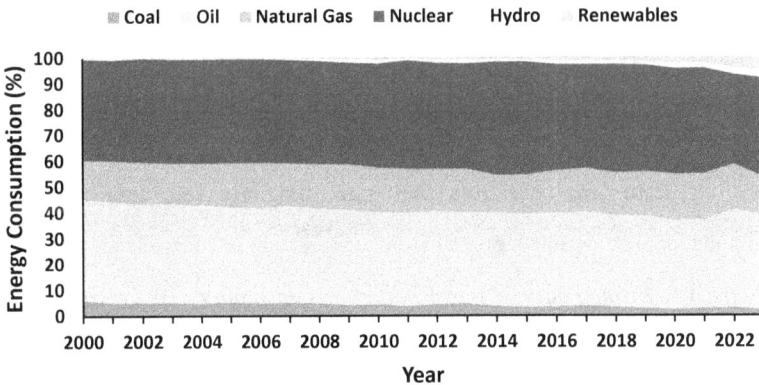

(b) Percentage consumption

Figure 4.3: France energy mix (2000 to 2023)

Source: EIA, EI

France is poor in terms of indigenous oil and natural gas reserves, which are estimated at 78.578 million bbl and 583 Bcf respectively.[12,13] With negligible natural gas production

---

[12] https://www.ogj.com/general-interest/economics-markets/article/14302481/global-oil-and-natural-gas-reserves-both-increase

[13] BGR Energiestudie 2023 - Daten und Entwicklungen der deutschen und globalen

domestically, France imports almost all its natural gas needs, as shown in Figure 4.4.

Although it is located in Western, not Eastern Mediterranean, France is a major actor in the Eastern Mediterranean natural gas affairs, to the extent that it officially joined EMGF. It is a staunch supporter of Cyprus and Greece against Turkey. There are historical and political reasons for France being one of the main opponents to Turkey joining the EU. In addition, there is an economic angle, as a French company, TotalEnergies, is a major player in exploration and possible production of natural gas in the Eastern Mediterranean Region.

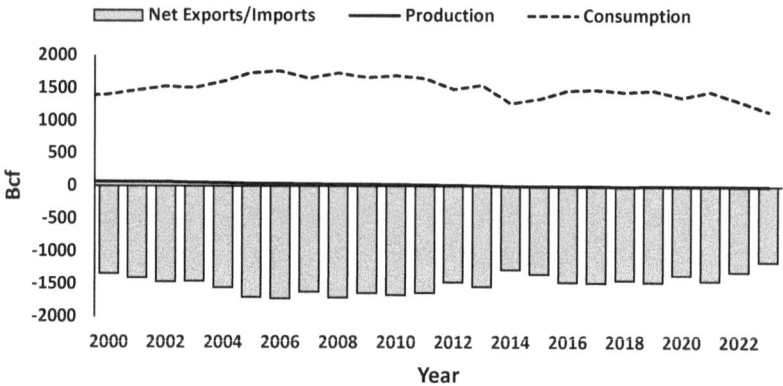

Figure 4.4: France natural gas production, consumption and trade (2000 to 2023)

Source: EIA, EI

France has an extensive internal domestic natural gas pipeline network/grid. Internationally, France is connected with some of its neighbours' natural gas pipelines. It imports significant volumes from Norway; to lesser extent Belgium and Germany; it has also connections with Switzerland and Luxembourg. There is no connection to the UK. Interestingly, it also has no connections with Italy, and limited connections to Spain, which limits both countries ambitions to be natural gas gateways to EU. This stance appears to be politically motivated. [As played out in the saga of the, now

Energieversorgung, potential additional resources 139 Tcf.

ditched, MidCat pipeline between Spain and France.[14,15,16] Since French pipelines do not have a direct effect on the Eastern Mediterranean Region natural gas dynamics, listing these pipelines is unnecessary for this book. However, France has 4 onshore and one offshore LNG regasification terminals with a total capacity of 28.3 mtpa. France historically imported LNG from Egypt, and after an interruption of several years, resumed this in 2022.

From a supply and demand point of view, France has little influence on the Eastern Mediterranean natural gas market, but as explained earlier, politically, it has significant influence via its companies and constant political interference in the region's affairs.

France has outlawed fracking and fossil fuel production in its territories and EEZ from 2040,[17] which will further diminish its influence and ambitions in the Eastern Mediterranean natural gas equation.

4.3   Iraq

Iraq is a large country both in terms of area and population, in the Persian/Arabian Gulf region. Its ancient civilisation goes back for thousands of years. Following decades of war with Iran, the USA and other Western countries, it has descended into chaos. Following Anglo-American invasion of 2003, the Western Alliance leaders disastrously failed to understand the complexity of the situation in Iraq and established a corrupt political system, mired by sectarian divisions. This effectively handed control of the country to neighbouring Iran. The Kurdish areas in the north are de-facto independent from the central government, with distinct

---

[14] https://www.euractiv.com/section/energy/news/france-trades-midcat-pipeline-for-an-already-controversial-new-project/

[15] Midi-Catalonia natural gas pipeline (MidCat) pipeline is a cancelled onshore pipeline project that was supposed to link Spain and France to allow for natural gas to flow from Spain to Europe via France. It faced numerous obstacle and delays prior to its cancellation. An alternative offshore project linking Barcelona to Marseille (H2Med; https://www.gem.wiki/H2Med_Pipeline) was announced in 2023 as an alternative pipeline.

[16] https://www.gem.wiki/Midi-Catalonia_Pipeline

[17] https://www.theguardian.com/environment/2017/dec/20/france-bans-fracking-and-oil-extraction-in-all-of-its-territories

economic, fiscal and foreign policies. Iraq is a member of the Arab League and a founding member of OPEC as well as a member of OAPEC. Iraq does not recognise Israel.

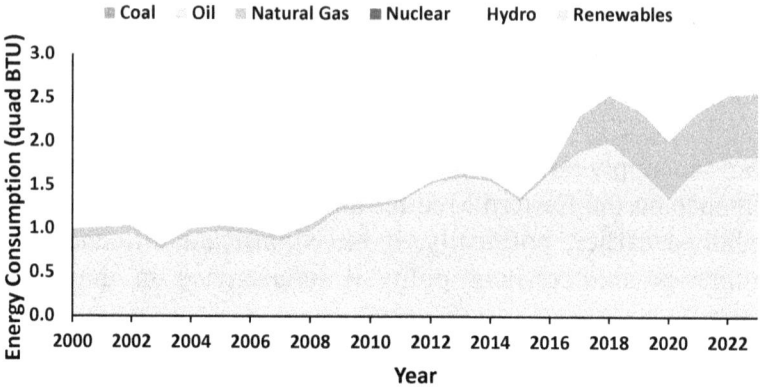

(a) Net consumption

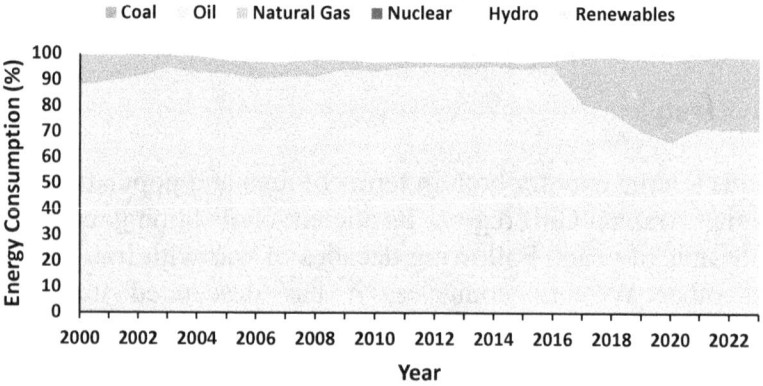

(b) Percentage consumption

Figure 4.5: Iraq energy mix (2000 to 2023)

Source: EIA, EI

Primary energy usage in Iraq fluctuates, although it appeared to have plateaued recently. Consumption is dominated by oil, but natural gas usage is growing steadily, and the energy sources are gradually diversifying. Figure 4.5 shows the energy mix and the trends, illustrating the growing role of natural gas in the energy mix.

Iraq has the fifth largest crude oil reserves in the world, estimated at 145.019 billion bbl.[18] It is a major oil producing and exporting country, as shown in Figure 4.6.

Iraq holds vast amounts of natural gas reserves, estimated at 131.156 Tcf at the end of 2023.[19,20] Its natural gas reserves places Iraq 12th in the world. Despite these reserves, Iraq's natural gas infrastructure and usage remain underdeveloped. The majority of the produced natural gas is either flared or re-injected to stimulate reservoirs and enhance oil recovery (EOR).

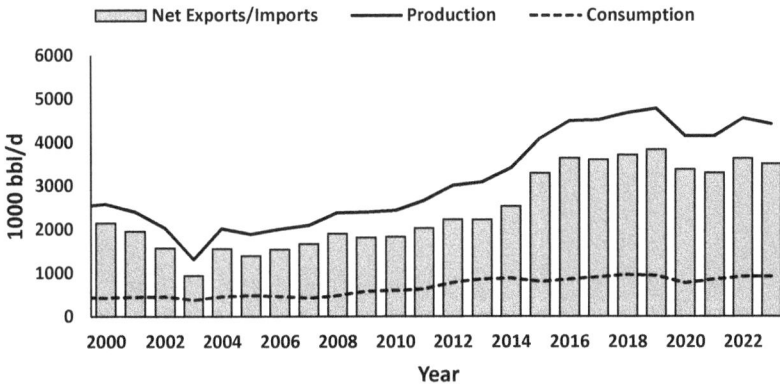

Figure 4.6: Iraq oil production, consumption and trade (2000 to 2023)
Source: EIA, EI

The flaring issue is endemic but several projects to capture and monetise the currently wasted natural gas are underway or planned. Major initiatives include the Basra Gas Company[21], that was established as a joint venture to reduce natural gas flaring from southern Iraqi oil fields and the TotalEnergies mega project, part of which aims to recover flared gas on three oil fields in order to

[18] https://www.ogj.com/general-interest/economics-markets/article/14302481/global-oil-and-natural-gas-reserves-both-increase
[19] https://www.ogj.com/general-interest/economics-markets/article/14302481/global-oil-and-natural-gas-reserves-both-increase
[20] BGR Energiestudie 2023 - Daten und Entwicklungen der deutschen und globalen Energieversorgung, potential additional resources 210 Tcf.
[21] https://www.shell.iq/en_iq/about-us/projects-and-sites/basrahgascompany.html

supply natural gas to power generation plants.[22]

Figure 4.7 shows Iraq's natural gas supply and demand, illustrating that, as its natural gas consumption was relatively small, Iraq was self-sufficient until 2016. It started becoming a net importer from 2017 despite the fact that it possesses a vast amount of natural gas reserves.

Despite being a net natural gas importer, Iraq is an observer of GECF.

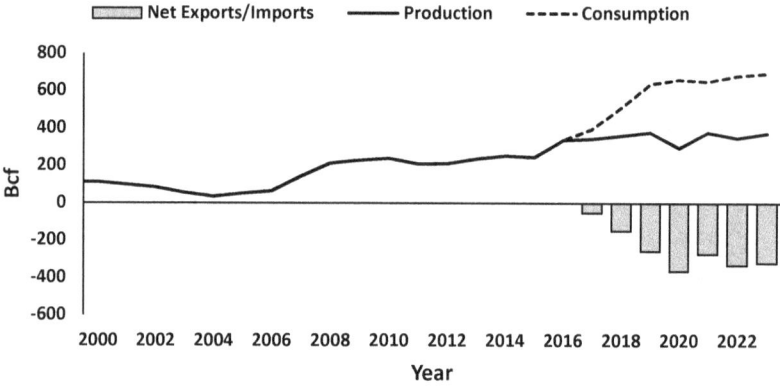

Figure 4.7: Iraq natural gas production, consumption, and trade (2000 to 2023)

Source: EIA, EI

Note that, prior to the 1990 invasion of Kuwait, Iraq exported natural gas to Kuwait to use in EOR.

Historically, Iraq is home to some of the oldest oil pipelines in the Middle East, some of which were significant influences on the drawing of the maps of the region.

Iraq has a limited natural gas domestic network/grid and few national pipelines. Iraq is linked, or plans to be linked, by natural gas pipelines to its neighbours. Recently, natural gas pipelines

---

[22] https://totalenergies.com/media/news/press-releases/iraq-agreement-between-iraq-and-totalenergies

were proposed as export pipelines or import pipelines, linking the Iraqi and Iranian natural gas grids.

Current operational pipelines are:

1. Iran-Iraq pipeline - Northern (See Section 4.4 – for details on Iran's pipelines).
2. Iran-Iraq pipeline - Southern (See Section 4.4 – for details on Iran's pipelines).

Other pipelines that are planned, proposed or cancelled include:

1. Iraq-Jordan pipeline, a cancelled natural gas pipeline that occasionally resurfaces. See below for more detailed discussion.
2. Iraq-Turkey pipeline, which is proposed project to export Iraqi natural gas to Turkey, linking to the Turkish pipe network and subsequently to Europe. In some drafts of the plan, it would extend to Ceyhan on the Mediterranean. It has been on the agenda for the last 40+ years (See Section 3.5 – for details on Turkey's pipelines).
3. Iraqi Kurdistan-Turkey pipeline, a proposed pipeline linking the Iraqi Kurdish natural gas network to the Turkish natural gas pipeline network. This would allow untapped natural gas reserves in Iraqi Kurdistan to be sold for Turkish domestic consumption or exported further to Europe. Its development is currently stalled due to Iraqi-Kurdish internal politics. (See Section 3.5 – for details on Turkey's pipelines).
4. Iraqi link to AGP, which never materialised. (See Section 3.8 – for details on Syria's pipelines).
5. Syria-Iraq pipeline, linking to the AGP to import Egyptian natural gas to Iraq. The idea is impractical. (See Section 3.8 – for details on Syria's pipelines).
6. Iraqi section of Iran-Iraq-Syria pipeline, which has been shelved after years of being discussed in the media. (See Section 3.8 – for details on Syria's pipelines).

7. Iraq-Kuwait pipeline,[23] proposed several times,[24] but abandoned.

Let's explore the story of one pipeline, in particular, as it encapsulates the story of Iraqi pipelines to other countries. The attempts at linking Iraq and Jordan via oil and gas pipelines is a long saga. Previous pipeline deals, dating from as far back as the 1980s, never materialised. In 2016 Jordan and Iraq announced another pipeline project[25] and tenders were sought in early 2017. The proposal was a double pipeline to export Iraqi crude oil and natural gas, via Jordan. This would provide Jordan with its oil and natural gas needs while the rest of the resources would be exported to Egypt and on to other international markets, via Aqaba. The routing of the pipe was shifted closer to the Iraqi Saudi border to avoid security risks in unstable regions of Iraq. The natural gas pipeline would be linked to the existing AGP to feed the Egyptian market.[26] As discussed earlier, the natural gas pipeline was cancelled one year later, in 2017 (see Section 3.7), while the oil pipeline is still planned to go ahead in 2024. (This is more than 40 years after the first proposal in the 1980s.)

Many Jordanians who oppose the natural gas supply deal with Israel have long argued (without proof) that this pipeline project would be a feasible alternative to dealing with Israel. However, given numerous aborted pipeline projects between the two countries, one cannot be optimistic that this project will ever materialise.

Then, for years, there was talk about an Iraqi LNG liquefaction project which would export LNG, onshore or FLNG.[27,28,29] All this talk never progressed and eventually, the project was

---

[23] https://www.gem.wiki/Iraq-Kuwait_Gas_Pipeline

[24] https://www.meed.com/kuwait-energy-to-tender-iraq-gas-pipeline-deals/

[25] Capacity is 1 million bpd oil and 258 MMscf/d natural gas.

[26] http://www.thearabweekly.com/Economy/6277/Iraq,-Jordan-pipelines-in-the-works

[27] https://2b1stconsulting.com/shell-and-south-gas-at-design-phase-on-first-iraq-lng-plant/

[28] https://lngjournal.com/index.php/latest-news-mainmenu-47/item/978-shell-explains-plans-for-flng-project-offshore-iraq

[29] https://www.iraq-businessnews.com/2011/01/26/mitsubishi-may-build-floating-lng-plant-off-iraq/#:~:text=Mitsubishi%20and%20Royal%20Dutch%20Shell,to%20the%20Wall%20Street%20Journal.

considered cancelled, but then, after being mentioned in statements by Iraqi officials regarding a proposed terminal at Al-Faw.[30], it was thought to be revived once again.[31] However, at the same time, there were announcements made regarding the construction of an LNG regasification project, also at Al-Faw.[32,33] Neither have come to fruition.[34] It is a total paradox.

At the moment, Iraq has no effects on Eastern Mediterranean natural gas market dynamics, as it is a net importer that secures its needs from Iran and recently, possibly, from Turkmenistan. Although Iraqi Kurdistan could sell some of its surplus natural gas to Turkey, as the planned pipeline to export it there has stalled to date, it is not possible to transport it to the market. The proposed pipeline is now a hostage to the political feud between the Iraqi federal government and the Kurdish regional government. Thus, in the short or medium term, Iraq's role is that of a transit state, allowing Iranian natural gas to reach new markets that can compete with natural gas from the Eastern Mediterranean.

4.4   Iran

Iran is the second largest country by area and the most populous in the Persian/Arabian Gulf region. The country is governed by an authoritarian Islamist regime. It is under sanctions imposed by the USA and several Western countries due to its efforts in developing nuclear weapons, despite its official denial. Iran is a founding member of OPEC, a member of GECF and it does not recognise Israel.

Primary energy usage in Iran continues to grow. Consumption is dominated by natural gas, whose usage is growing steadily and,

---

[30] https://oilprice.com/Energy/Natural-Gas/Will-Iraq-Finally-Increase-Natural-Gas-And-LNG-Production.html#:~:text=According%20to%20recent%20comments%20from,and%20sixth%20round%20of%20licensing.
[31] https://www.livemint.com/news/india/indias-lng-terminal-plans-in-iraq-face-visa-hurdles-11708425489794.html
[32] https://www.mees.com/2022/2/25/power-water/iraq-seeks-to-import-qatari-lng-but-how/0e586e60-964d-11ec-bde7-13ff8c48da79
[33] https://www.energyintel.com/0000018c-15ae-d3ae-af8f-7faead810000
[34] https://www.zawya.com/en/projects/construction/iraqs-proposed-lng-import-terminal-to-be-ready-in-2-years-report-ql10wkzg

apart from crude oil and natural gas, other energy sources are insignificant. Figure 4.8 shows the energy mix and the trends, illustrating the growing role of natural gas in the energy mix.

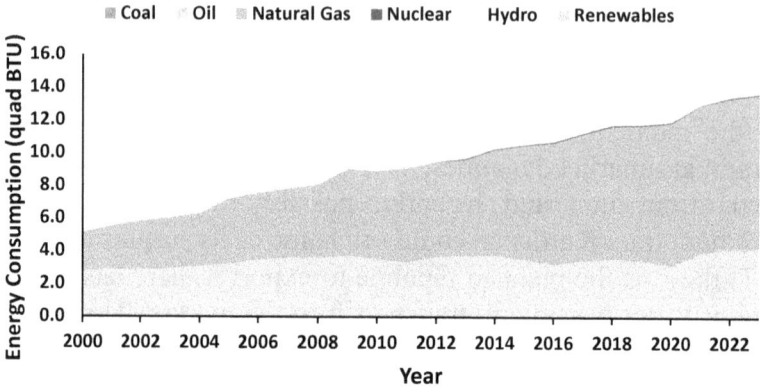

(a) Net consumption

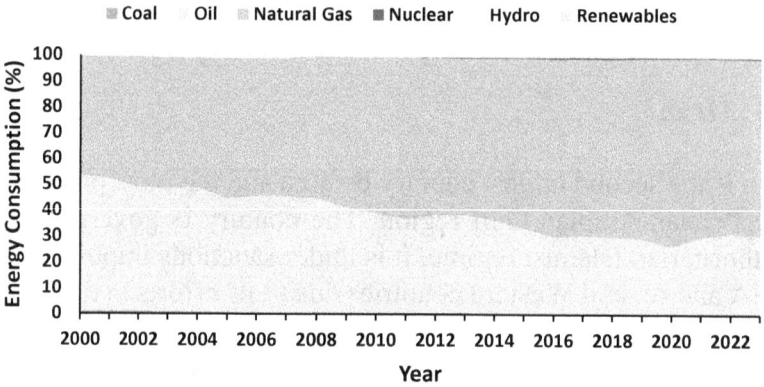

(b) Percentage consumption

Figure 4.8: Iran energy mix (2000 to 2023)

Source: EIA, EI

Iran is home to the fourth largest crude oil reserves in the world, estimated at 208.6 billion bbl at the end of 2023.[35] It is a major oil producing and exporting country, as shown in Figure 4.9. American sanctions have been in place since 1979, attempting to

---

[35] https://www.ogj.com/general-interest/economics-markets/article/14302481/global-oil-and-natural-gas-reserves-both-increase

curb and halt Iran's oil exports.

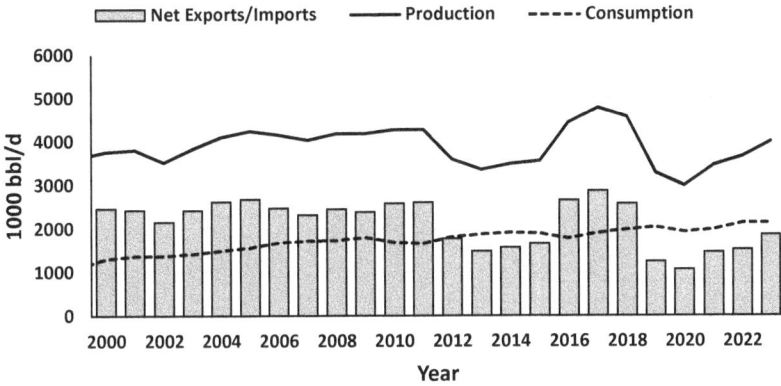

Figure 4.9: Iran oil production, consumption and trade (2000 to 2023)
Source: EIA, EI

Iran is home to the second largest reserves of natural gas in the world, after Russia. At the end of 2023, its vast amounts of natural gas reserves were estimated at 1200.252 Tcf.[36,37] However, despite these reserves, Iran's huge consumption of natural gas means that it has only small amounts to export as shown in Figure 4.10, hence its influence on global natural gas markets does not reflect the size of its reserves. Furthermore, despite its significant natural gas infrastructure, a lot of the reserves remain underdeveloped. At present their planned and/or proposed infrastructure projects are unlikely to proceed.

As a major source of natural gas and a potentially significant supplier, Iran has great ambitions to supply European, Iraqi and Lebanese customers, via several proposed pipelines that will pass through Syria and Turkey. These plans are part of Iran's vision to become a major global natural gas player and a transport hub to Central Asian natural gas also. At the moment, despite imposed American sanctions, Iran exports natural gas to several of its

---

[36] https://www.ogj.com/general-interest/economics-markets/article/14302481/global-oil-and-natural-gas-reserves-both-increase
[37] BGR Energiestudie 2023 - Daten und Entwicklungen der deutschen und globalen Energieversorgung, potential additional resources 350 Tcf.

neighbours. These exports are either exempt from the sanctions, such as to Iraq and Turkey, or despite the sanctions, they are tolerated e.g. swaps with Turkmenistan and Azerbaijan.

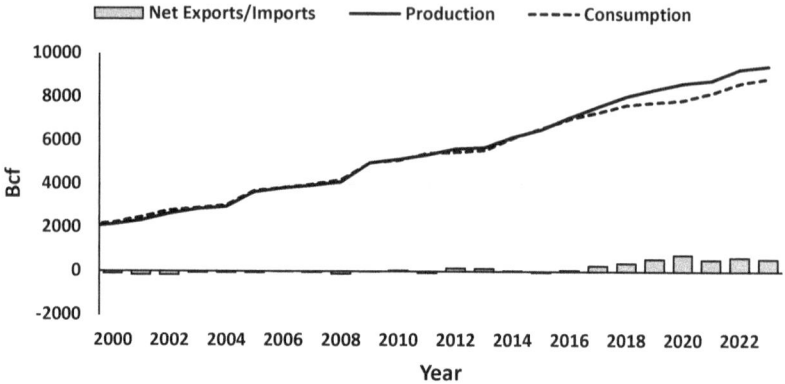

Figure 4.10: Iran natural gas production, consumption, and trade (2000 to 2023)

Source: EIA, EI

Iran has a substantial natural gas domestic network/grid and numerous intra national pipelines. This includes a series of large diameter pipelines, called the Iran Gas Trunkline (IGAT), that link natural gas sources in the South to consumption centres. Furthermore, the country is linked, or plans to be linked, by natural gas pipelines to most of its neighbours. Currently, several natural gas pipelines are used to export or swap production. These cross-border pipelines are listed below:

1. Iran-Iraq northern onshore pipeline,[38] that links the Iranian natural gas grid to the Iraqi natural gas grid supplying natural gas to Iraqi power stations near Baghdad. It started operating in 2017.[39]

2. Iran-Iraq southern onshore pipeline,[40] that links the Iranian natural gas grid to the Iraqi natural gas grid supplying natural gas to Iraqi power stations near

---

[38] Pipeline is 135 km, diameter is 48"; annual capacity is 12.8 bcm (447 Bcf).
[39] https://indianexpress.com/article/world/iran-begins-sending-gas-to-iraq-under-major-deal-4717091/
[40] Pipeline is 141 km, diameter is 48"; annual capacity is 9.1 bcm (318 Bcf).

Basra. It started operating in 2018.[41]

3. Tabriz–Ankara operational onshore pipeline,[42] linking the Iranian natural gas network to the Turkish natural gas network. (See Section 3.5 – for details on Turkey's pipelines).

4. Hajiqabul–Astara–Abadan gas onshore pipeline,[43] linking Azerbaijan and Iran which goes back to the Soviet era. It is one of several natural gas pipelines between the two countries that allows natural gas swap deals, that also involves Turkmen natural gas.

5. Rasht-Chelavand gas pipeline,[44] is a newly completed natural gas pipeline linking Azerbaijan and Iran. It allows swaps between the two countries and Turkmenistan. It is the shortest route for Russian natural gas to reach Iran.

6. Iran-Nakhchivan operational onshore pipeline, supplying the Azerbaijani exclave of Nakhchivan with Iranian natural gas, as part of a swap deal with Azerbaijan.[45,46]

7. Iran-Armenia operational natural gas pipeline,[47] supplying Armenia with Iranian natural gas.

8. Korpeje-Kordkuy gas onshore pipeline,[48] is an operating pipeline to transport Turkmen natural gas to Iran and possibly beyond.[49]

9. Dauletabad-Sarakhs-Khangiran gas pipeline,[50] is an operating onshore pipeline to transport Turkmen natural gas to Iran and possibly beyond.[51]

---

[41] https://www.iraq-businessnews.com/2019/05/15/iran-boosting-gas-exports-to-iraq/

[42] https://www.gem.wiki/Tabriz-Ankara_Pipeline

[43] https://www.gem.wiki/Hajiqabul%E2%80%93Astara%E2%80%93Abadan_Gas_Pipeline

[44] https://www.newscentralasia.net/2023/02/16/remaining-section-of-rasht-chelavend-gas-pipeline-will-be-completed-in-2-3-months-it-will-add-5-5-bcm-to-gas-transportation-capacity-between-iran-and-azerbaijan/

[45] https://www.aa.com.tr/en/world/baku-signs-deal-with-tehran-for-transportation-lines-to-nakhchivan-via-iran/3014861

[46] https://www.oxfordenergy.org/wpcms/wp-content/uploads/2013/06/NG-78.pdf

[47] https://www.gem.wiki/Iran%E2%80%93Armenia_gas_pipeline

[48] https://www.gem.wiki/Korpeje-Kordkuy_Gas_Pipeline

[49] https://eurasianet.org/turkmenistan-turkey-talk-gas-transit-but-iran-factor-muddies-picture#:~:text=There%20are%20two%20pipelines%20able,cubic%20meter%20Iran%2DTurkey%20pipeline.

[50] https://www.gem.wiki/Dauletabad-Sarakhs-Khangiran_Gas_Pipeline

[51] Ibid.

Other pipelines are planned, proposed, or cancelled. These include:

1. Iran-Iraq-Syria pipeline,[52] dubbed the 'Islamic Pipeline' or 'Friendship Pipeline', is a shelved onshore pipeline, linking Iran to Syria via Iraq, to export Iranian natural gas to Syria and Lebanon, and possibly Europe. (See Section 3.8 – for details on Syria's pipelines). Iran is trying to revive the project.[53]
2. Iran Kuwait pipeline, cancelled offshore pipeline to supply Kuwait with Iranian natural gas[54]
3. Persian Gas Pipeline, (also known as Pars Pipeline, Iran–Europe pipeline, or Iran–Turkey-Europe (ITE) pipeline),[55,] a proposed onshore pipeline to transport Iranian natural gas to Europe via Turkey. (See Section 3.5 – for details on Turkey's pipelines).
4. Iran-Oman pipeline, a proposed offshore pipeline to supply natural gas to Oman's LNG terminals. Both countries are committed to it, but it is yet to materialise.[56]
5. Iran-Bahrain pipeline, a cancelled offshore pipeline to supply Bahrain with Iranian natural gas.[57]
6. Iran-UAE pipeline, a constructed, yet never operated, pipeline linking to Sharjah, to supply the UAE with Iranian natural gas.[58]
7. Iran-Pakistan pipeline, under construction onshore pipeline, with the Iranian part complete.[59]
8. Iran-Pakistan-India pipeline, cancelled onshore project, dubbed the peace project.[60] The Iran-Pakistan pipeline above is part of this original project.
9. In addition, there is talk about a proposed natural gas pipeline from Russia, utilising the Iranian network and then

---

[52] https://www.gem.wiki/Iran%E2%80%93Iraq%E2%80%93Syria_pipeline
[53] https://oilprice.com/Energy/Energy-General/Why-Iran--Is-Desperate-To-Finish-This-Pipeline.html
[54] https://www.meed.com/iran-agrees-kuwait-gas-export-deal/
[55] https://www.gem.wiki/Persian_Gas_Pipeline
[56] https://www.gem.wiki/Iran-Oman_Gas_Pipeline
[57] https://www.oxfordenergy.org/wpcms/wp-content/uploads/2013/06/NG-78.pdf
[58] Ibid.
[59] https://www.gem.wiki/Iran-Pakistan_Pipeline
[60] https://www.gem.wiki/Iran-Pakistan-India_Pipeline

routed offshore to Pakistan. The exact route has not been determined and the project remains speculative.[61]

After decades of plans, false starts and more than 6 aborted LNG liquefaction projects,[62] the first LNG liquefaction project in Iran is finally underway at Asaluyeh in South Iran, but at a capacity of 1.5 mtpa, this makes it, if completed, a minor player in the LNG market.[63] Even if the US sanctions were to be lifted, the development of further LNG liquefaction capacity will be slow, problematic, and finding markets for the produced natural gas would be difficult.

Iran involves itself constantly in the politics of the Eastern Mediterranean region, either directly, via its military presence in Syria, or indirectly, via its proxies in Lebanon, Iraq and Yemen. Iran's influence on world events is clear and has recently been illustrated by the threats to global trade routes. An example, that includes LNG trade, is where the Houthis of Yemen, an Iranian proxy, have been disrupting the shipping flow in the Red Sea with attacks on ships off Yemini shores, in response to the Israeli attacks on Gaza. Note, that its military drone and missile attack on Israel in 2024 was ineffective.

Finally, despite the huge potential for Iran to be a major natural gas source for the world, the stranglehold of tightening American sanctions is an obstacle that continues to curtail this potential and limit its natural gas developments.

4.5   Qatar

This is a small Arab emirate, located in the Eastern part of Arabia that punches above its weight due to its wealth and the size of its natural gas reserves. Moreover, its support of Al-Jazeera, the most influential news channel in the Arab world, means that Qatar has an enormous influence in the Arab world. Qatar was a former

[61] https://www.gem.wiki/Off-Shore_Gas_Pipeline
[62] https://www.oxfordenergy.org/wpcms/wp-content/uploads/2013/06/NG-78.pdf
[63] https://www.spglobal.com/commodityinsights/en/market-insights/latest-news/lng/012524-interview-iran-aims-for-15-mil-mtyear-lng-output-in-2026-deputy-minister

member of OPEC.[64] It is a member of GECF, the Arab League and of OAPEC. Qatar does not recognise Israel.

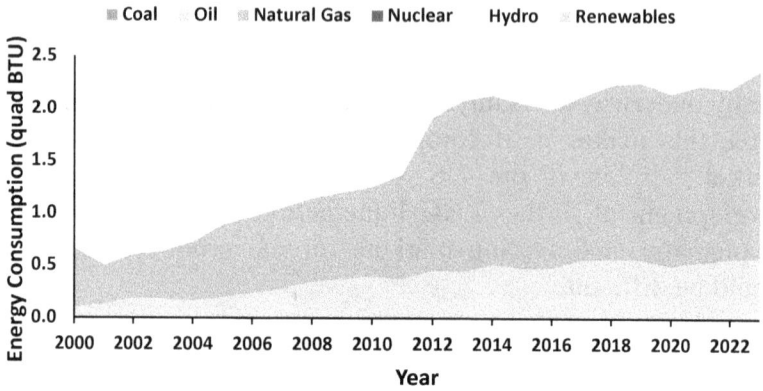

(a) Net consumption

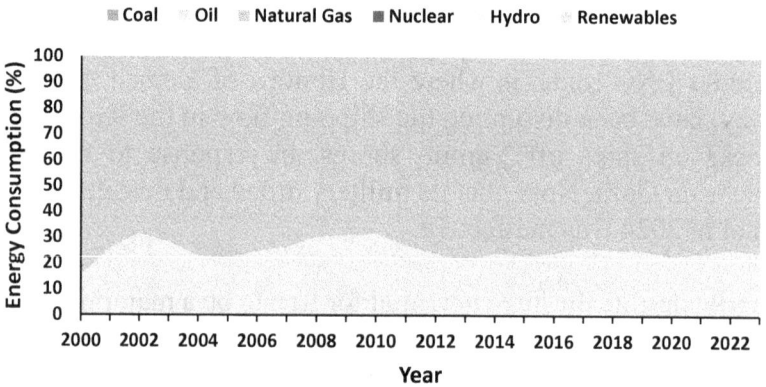

(b) Percentage consumption

Figure 4.11: Qatar energy mix (2000 to 2023)

Source: EIA, EI

Primary energy usage in Qatar has plateaued. Consumption is dominated by natural gas, and apart from crude oil and natural gas, other energy sources are insignificant. Figure 4.11 shows the energy mix and the trends, illustrating the dominant role of natural gas in the energy mix.

---

[64] Qatar withdrew from OPEC in 2019 for political reasons and to avoid the threat of NOPEC law. Although it cited its reduced oil production as a reason for withdrawal, it in fact produces more crude oil and NGL than several OPEC members, including Libya, and Algeria.

Qatar has substantial crude oil reserves, estimated at 25.244 billion bbl at the end of 2023.[65] However, it is a medium-sized oil producing and exporting country, as shown in Figure 4.12.

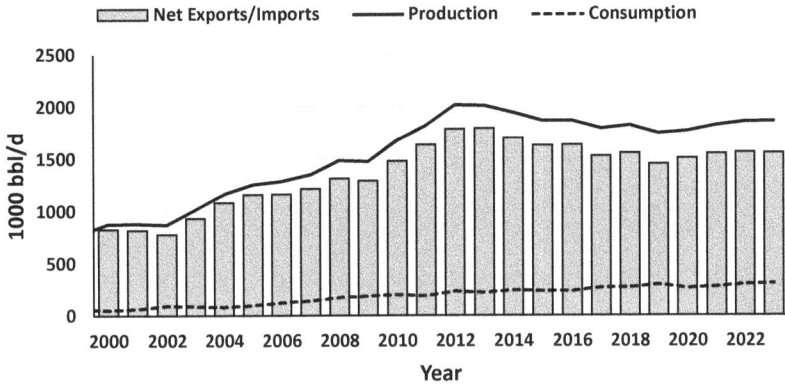

Figure 4.12: Qatar oil production, consumption and trade (2000 to 2023)
Source: EIA, EI

Qatar is home to the third largest reserves of natural gas in the world, after Russia and Iran. Its vast amounts of natural gas reserves are estimated at 842.627 Tcf, at the end of 2023.[66,67] Unlike Iran, Qatar has embarked on major developments of its vast reserves, making it one of the three major exporters of LNG in the world, with total capacity of 77.4 mtpa (the others being USA and Australia). Currently, Qatar is expanding its LNG capacity by an additional 48 mpta, aiming to maintain its dominant position in the market. Figure 4.13 shows Qatar's natural gas supply and demand, demonstrating the large surplus available to export.

Qatar has limited natural gas export pipelines, with only the Qatar-UAE pipeline working. This is part of the Dolphin project that also supplies natural gas to Oman.

---

[65] https://www.ogj.com/general-interest/economics-markets/article/14302481/global-oil-and-natural-gas-reserves-both-increase
[66] Ibid.
[67] BGR Energiestudie 2023 - Daten und Entwicklungen der deutschen und globalen Energieversorgung, potential additional resources 70 Tcf.

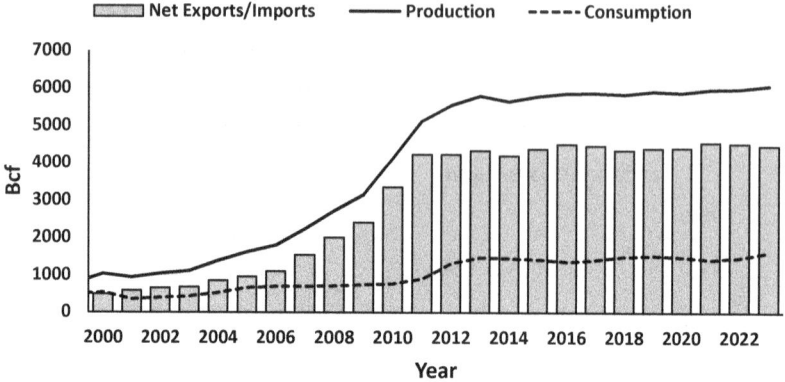

Figure 4.13: Qatar natural gas production, consumption and trade (2000 to 2023)

Source: EIA, EI

Planned or proposed pipelines include:

1. Qatar-Kuwait pipeline, and subsequent expansion to Iraq. This pipeline never materialised because Saudi Arabia did not allow passage through its EEZ. There were ambitions to extend this pipeline to Bahrain, and also from Iraq, to connect to the Nabucco pipeline to Europe (See Section 3.5 – for details on Turkey's pipelines).

2. Extending the Qatar-UAE and Oman pipeline, with proposed extensions to Pakistan and India, eastwards.

3. The media reported the possibility of a natural gas pipeline from Qatar to Europe, via Turkey, Syria and Jordan. Conspiracy theorists have suggested that this pipeline was the reason for the Syrian civil war. These theories have been totally debunked. The pipeline is not only uneconomic and unfeasible, (transport costs will erode any price advantages), but Saudi Arabia will never allow the routing, especially after it previously rejected a pipeline to fellow GCC member Kuwait.

As mentioned earlier, Qatar is one of the 'top three', in terms of LNG exports, in the last decade, competing with Australia and the

USA as a top LNG exporter. See Figure 4.14. Nowadays the USA in first place.[68]

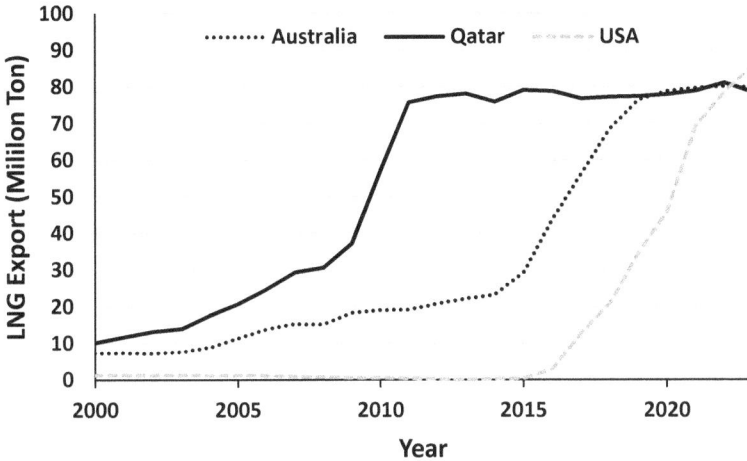

Figure 4.14: LNG exports from Qatar, Australia and USA (2000 to 2023)

Source: EIA

As we can see, Qatar is a major global influencer of the natural gas markets, which of course affects the natural gas market dynamics in the Eastern Mediterranean. In addition, QatarEnergy is increasingly investing in acquiring shares in concessions and farming in the Eastern Mediterranean blocks of Egypt,[69] Lebanon[70] and Cyprus.[71]

4.6 Algeria

Algeria is the largest African and Arab country by area and is located on the southern shores of the Western Mediterranean. The country's government still lives in the nationalist and socialist past, and its policies have hindered developing the country to its

---

[68] https://www.ft.com/content/2d78c322-ac8a-4595-add6-cf182d1712d7?desktop=true&segmentId=dd5c99e9-30be-ddd0-c634-ff3a0c2b738f#myft:%E2%80%A6

[69] https://jpt.spe.org/qatarenergy-grows-east-med-presence-with-exxonmobil-farm-in-offshore-egypt

[70] https://www.offshore-technology.com/news/qatarenergy-consortium-lebanon/

[71] https://www.offshore-technology.com/news/qatarenergy-exxonmobil-offshore-block-cyprus/

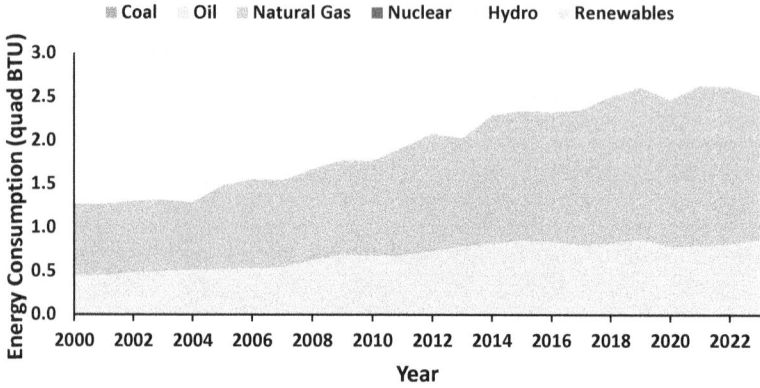

(a) Net consumption

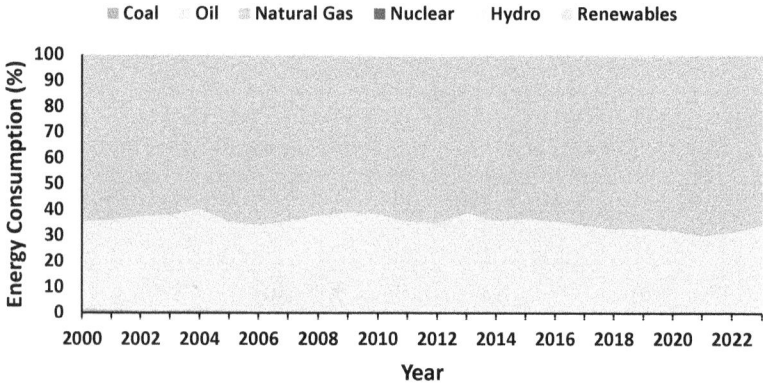

(b) Percentage consumption

Figure 4.15: Algeria energy mix (2000 to 2023)

Source: EIA, EI

potential. Algeria's relationship with France remains uneasy, due to their history of being a colony for 132 years They only achieved independence by overthrowing their French colonisers in 1962. It also has a tense relationship with its main neighbour Morocco. Algeria is a member of the Arab League, the GECF, as well both OPEC and OAPEC. It does not recognise Israel.

Primary energy usage in Algeria appears to have plateaued. Consumption is dominated by natural gas, and apart from crude oil and natural gas, other energy sources are insignificant. Figure 4.15 shows the energy mix and the trends, illustrating the major role of natural gas in the energy mix.

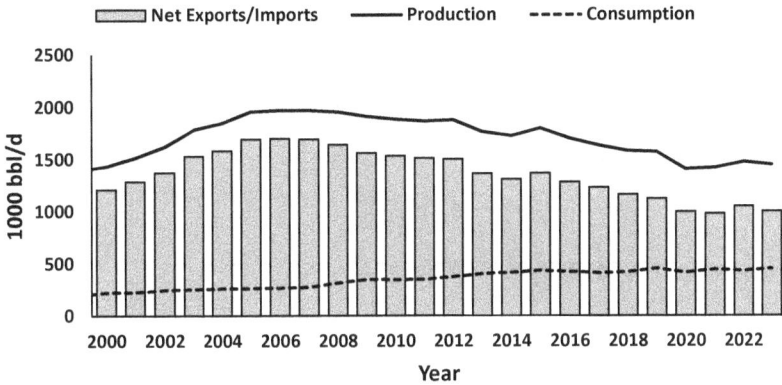

Figure 4.16: Algeria oil production, consumption and trade (2000 to 2023)
Source: EIA, EI

Algeria has substantial crude oil reserves, estimated at 12.2 billion bbl.[72] It is ranked third in Africa and 16th globally. However, it is a medium-sized oil producing and exporting country, as shown in Figure 4.16.

Algeria has substantial natural gas reserves, placing it second in Africa and 11th globally. At the end of 2023, reserves were estimated as 159.054 Tcf.[73,74] It also has a huge 707 Tcf unconventional, technically recoverable, shale gas resources which are yet to be exploited and developed.[75] It is a major natural gas player on the global market. It is a major exporter as shown in Figure 4.17 and supplies Europe with around 12% of its natural gas needs.[76] In addition, it is the home of the first LNG liquefaction

---

[72] https://www.ogj.com/general-interest/economics-markets/article/14302481/global-oil-and-natural-gas-reserves-both-increase
[73] https://www.ogj.com/general-interest/economics-markets/article/14302481/global-oil-and-natural-gas-reserves-both-increase
[74] BGR Energiestudie 2023 - Daten und Entwicklungen der deutschen und globalen Energieversorgung, potential additional resources 956 Tcf.
[75] https://www.energyintel.com/0000018d-d105-d9ab-adff-d32d72b80000#:~:text=With%20some%20707%20trillion%20cubic,false%20starts%20in%20the%20past.
[76] https://www.icis.com/explore/resources/news/2024/03/20/10982362/algeria-eyes-greater-role-

terminal in the world, which started operation in 1965[77] and remains a major LNG exporter today, with 25.3 mtpa capacity. Its influence in that market is diminishing. In the Eastern Mediterranean mix, it plays modest role. It is a major LNG supplier to Turkey; it also sends a small amount of LNG loads to Jordan and Egypt.

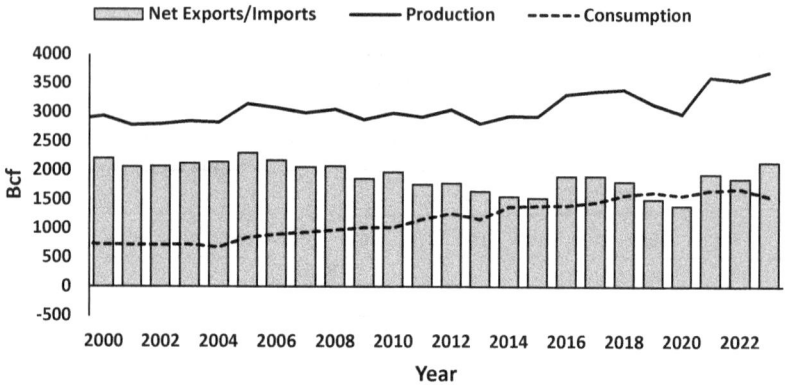

Figure 4.17: Algeria natural gas production, consumption and trade (2000 to 2023)

Source: EIA, EI

Algeria has substantial natural gas domestic network/grid and numerous intra-national pipelines. Furthermore, the country is linked by natural gas pipelines to some of its neighbours and offshore, further afield. Currently, Algeria uses pipelines as major route to export natural gas to the EU. Since the Russian invasion of Ukraine, with alternative natural gas sources for Europe being sought, Algeria's importance has increased significantly.

There are currently several pipelines connecting Algeria to Europe, either directly under the sea, or transiting via Tunisia or Morocco. These lines are:

1. Medgaz, connecting Algeria to Spain, via an underwater direct route.

---

as-europe-s-gas-supplier/
[77] Azrew-Skikda LNG

2. Maghreb–Europe Gas Pipeline (MEG), connecting Algeria to Spain and Portugal via Morocco. The use of this export pipeline has been suspended due to political disagreements between Algeria and Morocco, although part of it has been reversed, sending natural gas from Spain (originally imported as LNG) to Morocco.
3. Trans-Mediterranean pipeline, connecting Algeria to Italy and Slovenia, via Tunisia.
4. Nigeria-Algeria pipeline, a strategic trans-African pipeline which is in the planning stages. It will connect Nigeria to Algeria via Niger and then use current existing Algerian pipelines to transport gas to Europe. Note, this pipeline is in direct competition to two other pipeline projects. The first connecting Nigeria, via an offshore pipeline running parallel to the West Africa coast and Morocco's coast to Europe,[78] the second is the Libyan proposal to reroute the Nigerian pipeline, via Libya instead of Algeria (see Section 3.10 – for details on Libya's pipelines).
5. GALSI pipeline, planned to connect Algeria and Italy directly via submarine pipeline, but this has been shelved.

With its geographical location, current and potential natural gas reserves, Algeria could be a secure source for supplying Europe with its natural gas needs. However, to reach its full potential, it would require vast investments and the liberalisation of its laws in order to allow easier access for international companies to work there. Once it creates these optimal conditions for the development of its natural gas industry, it could be a formidable competitor to natural gas supply from the Eastern Mediterranean region to Europe, since it is already linked to the EU by several pipelines. Although with Europe's aim to diversify resources to secure its supplies, Libya, Cyprus and/or Israel may have a place in the European market. Of course, if Algeria is not to make the necessary changes, then the opportunity will be lost.

---

[78] https://www.pipeline-journal.net/news/nigeria-morocco-gas-pipeline-project-moves-forward-northern-section-survey-begins

4.7 Russia

Russia is the world's largest country by area and, for decades, as the dominant Republic in the Soviet Union, it was the world's second superpower. This position weakened following the collapse of the Soviet Union in 1991, although it remains a major nuclear power, with military bases in the Eastern Mediterranean in Syria

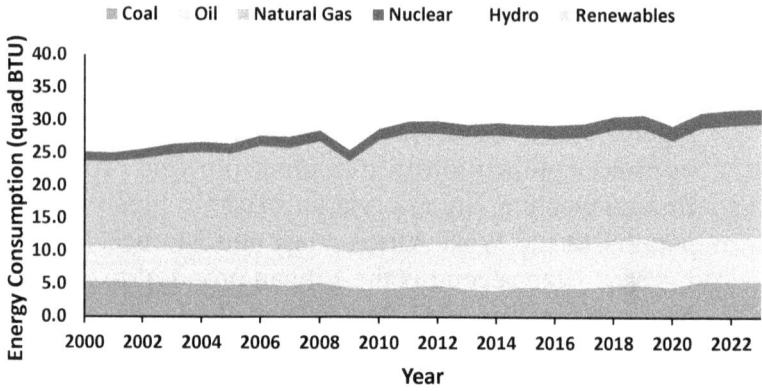

(a) Net consumption

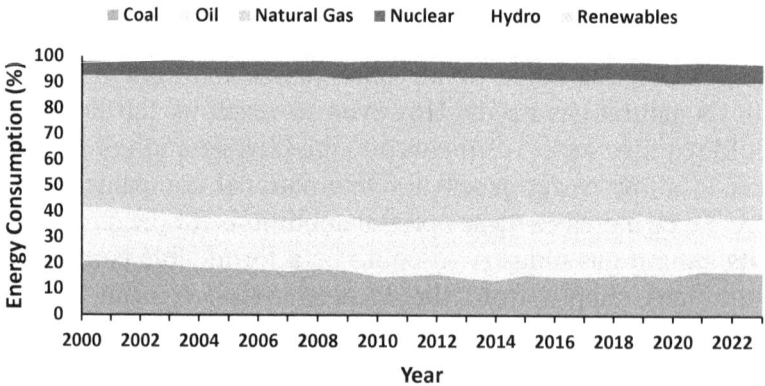

(b) Percentage consumption

Figure 4.18: Russia energy mix (2000 to 2023)

Source: EIA, EI

and an unofficial presence in Libya. Since 2022, following the invasion of Ukraine, Russia is under severe (yet ineffective)

Western sanctions, which are not enforced by the rest of the world, including Turkey and Israel, who simply ignore them.

The country is a member of GECF, and although not a member of OPEC, is an essential member of OPEC+, attempting to manipulate the price of crude oil.

Primary energy usage in Russia continues to grow. Consumption is diverse, but natural gas has the largest portion. Figure 4.18 shows the energy mix and the trends, illustrating the growing role of natural gas in the energy mix.

Russia has substantial crude oil reserves, estimated at 80 billion bbl (ranked eighth).[79] It is one of the top crude oil producing and exporting countries, as shown in Figure 4.19. Despite the tough international sanctions affecting Russian oil, following the Ukrainian invasion, the level of production and exports did not drop as substantially as anticipated by Western nations, since many other countries ignored the sanctions and continued to purchase Russian oil.

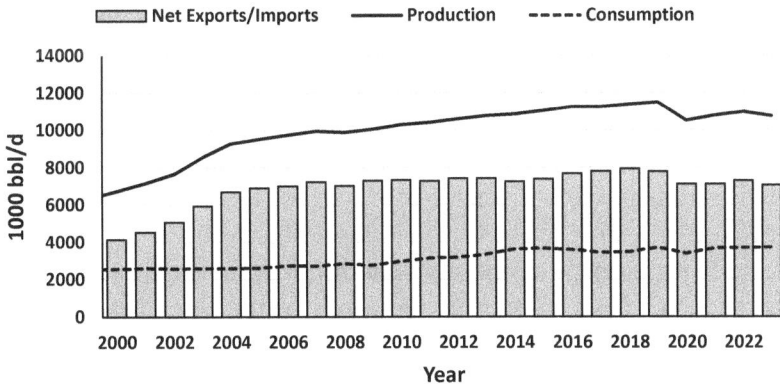

Figure 4.19: Russia oil production, consumption, and trade (2000 to 2023)

Source: EIA, EI

---

[79] https://www.ogj.com/general-interest/economics-markets/article/14302481/global-oil-and-natural-gas-reserves-both-increase

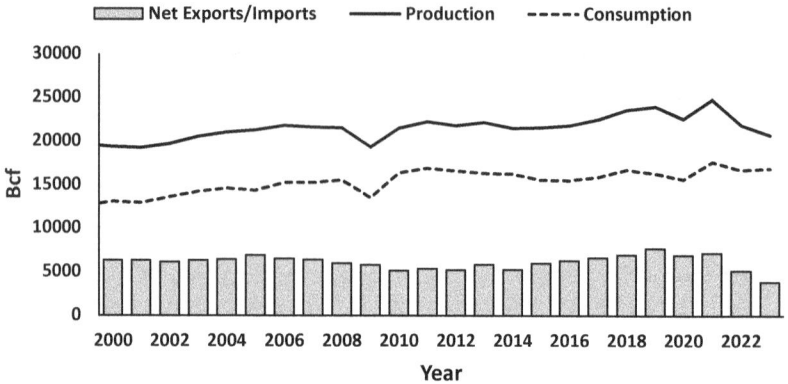

Figure 4.20: Russia natural gas production, consumption and trade (2000 to 2023)

Source: EIA, EI

Russia has the world's largest reserves of natural gas. At the end of 2023, its vast amounts of natural gas reserves were estimated at 1688.228 Tcf.[80,81] It is the world second largest producer and exporter (see Figure 4.20). Despite Western sanctions, the drop in both production and exports was less than anticipated.

Russia has an extensive internal domestic natural gas pipeline network/grid. In addition, Russia has the longest natural gas pipelines in the world and is connected to almost all of its neighbours. Listing these pipelines is outside the scope of this book. The important connections affecting the Eastern Mediterranean were discussed earlier (see Section 3.5 and 3.8).

Besides its extensive pipeline network, Russia is exporting natural gas as LNG from numerous terminals in the Far East, the Arctic and the Baltic Sea. It has total capacity of 37 mtpa. These exports were hardly hit (in fact grew) by Western sanctions and to the dismay of Western nations, Russia started up its latest LNG terminal, Arctic LNG operated by Novatek, in 2024, defying all

---

[80] https://www.ogj.com/general-interest/economics-markets/article/14302481/global-oil-and-natural-gas-reserves-both-increase.
[81] BGR Energiestudie 2023 - Daten und Entwicklungen der deutschen und globalen Energieversorgung, potential additional resources 5082 Tcf.

the sanctions.

Although Russia has no direct influence on the natural gas market and development in the Eastern Mediterranean, in fact the dynamics of Russia-EU natural gas trade has a huge impact on the trajectory of natural gas development in the Eastern Mediterranean since both natural gas producing areas compete to supply same market. Following the elimination of Russian natural gas from the European equation, natural gas from the Eastern Mediterranean has the potential to fill part of the gap. [82] Also despite the sanctions, Russian natural gas can still find its way to Europe via Turkey, either sent directly from Russia or indirectly, via Caspian and future Central Asian routes.

Furthermore, following Russia's invasion of Ukraine, Russian companies farmed-out and disposed of some of their concessions of exploration blocks in the Eastern Mediterranean, particularly in Lebanon. At the moment their involvement continues in Egypt and Syria.

4.8   USA

The world's number one superpower has always maintained a significant presence in the Eastern Mediterranean Region. It has several military bases in the region. Its foreign policies have shaped the region for decades and recent attempts of some American administrations to disengage from the region have been reversed following the Russian invasion of Ukraine in 2022 and the Gaza War of 2023-present.

Primary energy usage in USA appears to have plateaued. Consumption sources are diverse, but fossil fuels continue to dominate. Figure 4.21 shows the energy mix and the trends, illustrating the significant role of natural gas in the energy mix.

The unconventional oil and natural gas revolution in the last decade has redrawn the map of oil and natural gas trade, altering

---

[82] Some EU countries have exemptions to import Russian natural gas, and LNG is not sanctioned.

their market dynamics beyond recognition.

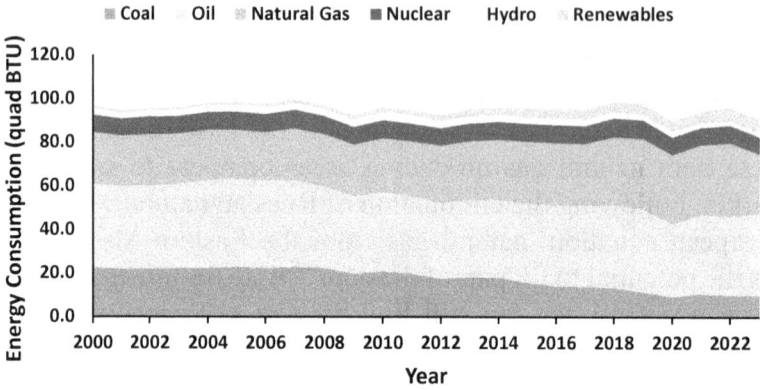

(a) Net consumption

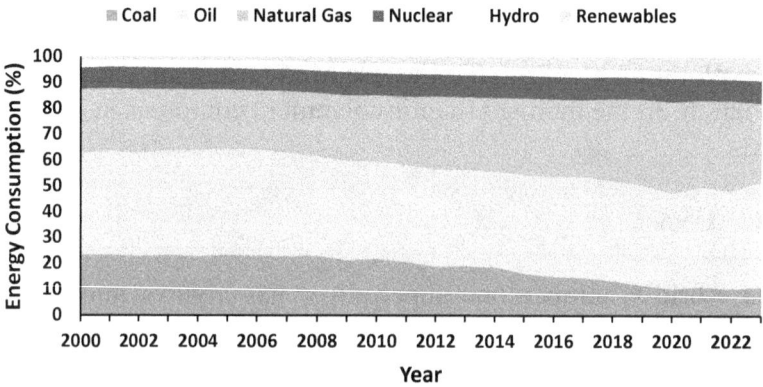

(b) Percentage consumption

Figure 4.21: USA energy mix (2000 to 2023)

Source: EIA, EI

In a matter of few years the US transformed itself from a net importer to a net exporter for both crude oil and natural gas. See Figures 4.22 and 4.23.

In terms of crude oil, the USA has substantial crude oil reserves (including a significant proportion of unconventional resources that are now classified as reserves). By the end of 2023, these reserves placed the country ninth globally and are estimated at

74.417 billion bbl.[83] In 2024 the USA is the top crude oil producing country, albeit it is a minor net exporting country as shown in Figure 4.22.

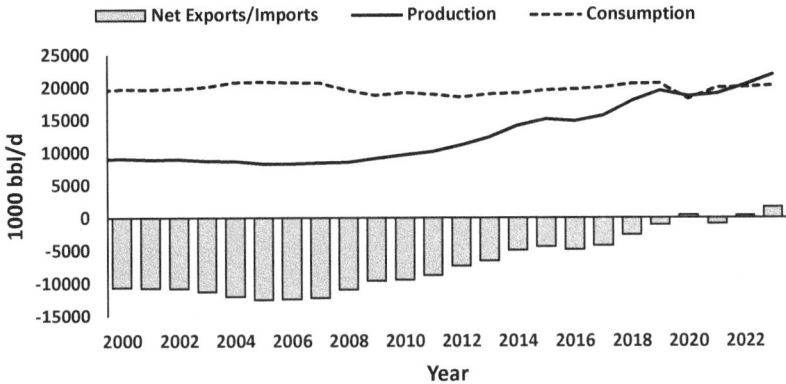

Figure 4.22: USA oil production, consumption and trade (2000 to 2023)
Source: EIA, EI

Similarly in terms of natural gas, the USA has substantial natural gas reserves (including a significant proportion of unconventional resources that are now classified as reserves) placing the country fourth globally. These reserves were estimated at 615.125 Tcf at the end of 2023.[84,85,86] Currently, the country is the top producer exporter (see Figure 4.23).

The USA has an extensive internal domestic natural gas pipeline network/grid. Although the country has extensive networks of intra-state and country pipelines, its international links to Canada and Mexico are limited. Due to their geographical location, far from the Eastern Mediterranean, listing these pipelines is outside the scope of this book.

---

[83] https://www.ogj.com/general-interest/economics-markets/article/14302481/global-oil-and-natural-gas-reserves-both-increase
[84] Ibid.
[85] BGR Energiestudie 2023 - Daten und Entwicklungen der deutschen und globalen Energieversorgung, potential additional resources 3800 Tcf.
[86] According to EIA latest published numbers, reserved in 2022 were 691.025 Tcf - https://www.eia.gov/naturalgas/crudeoilreserves/index.php

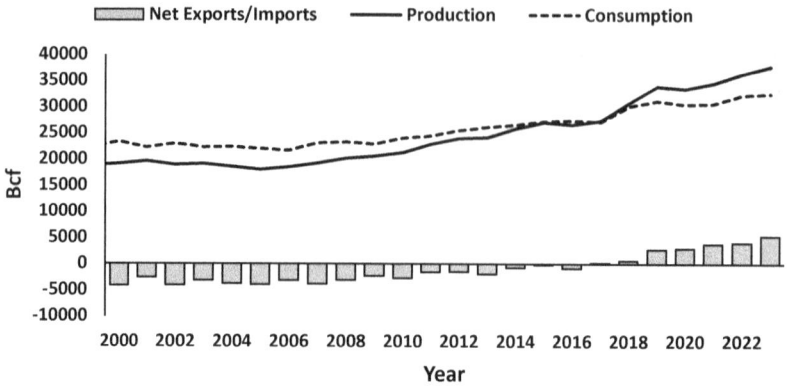

Figure 4.23: USA natural gas production, consumption and trade (2000 to 2023)

Source: EIA, EI

After the USA turned into a net natural gas exporter, a wave of LNG export terminals were constructed and almost all previous import terminals on the Gulf Coast were converted to export terminals. At the start of 2024, the total liquefaction capacity was 83.8 mtpa, the largest in the world. At the time of writing, more terminals are under construction (84.1 mtpa) or proposed (254 mtpa), and the temporary pause by the Biden administration to give permits to further developments has been lifted but is still being contested in the courts.

This aggressive expansion has created fierce competition for the natural gas producers in the world, as these US independent producers aggressively marketed their products and offered them in the spot market.

Indeed, the dynamics of USA-EU natural gas trade has an enormous impact on the trajectory of natural gas development in the Eastern Mediterranean since the natural gas producing areas compete to supply same market. In this case, pure LNG versus a mix of pipeline natural gas and LNG.

Furthermore, American companies are also major players in the Eastern Mediterranean. Chevron controls the Israeli natural gas export market, while both Chevron and ExxonMobil are involved in Egypt and Cyprus. Thus, promoting Israeli natural gas as a potential supply source to Europe can be influenced by the US economic interests, both supporting American companies, while geopolitically fulfilling American pledges providing support to Israel.

It is worth noting though, that the rapid expansion in energy demand by data centres and artificial intelligence (AI) adoption, may influence the future US natural gas export potential. The rapidly expanding demand may divert some production to the domestic market, while the US administration may be also tempted to restrict exports for geopolitical and economic reasons.

4.9   Others

Other countries policies close to the region or further away also influence the natural gas dynamics in the Eastern Mediterranean region. These are summarised below:

- **Malta** has EEZ disputes with Libya, Tunisia and Italy that may lead to future disputes if natural gas is discovered in these areas. Currently Malta is a minor natural gas importer since its market size is small, but as an EU member, its natural gas policies are governed by EU overall policy. It imports natural gas via a tiny 0.5 mtpa FSRU LNG regasification terminal.[87] Thus, despite being a neighbouring country to the region, Malta has trivial influence on the Eastern Mediterranean natural gas market dynamics.

- **Tunisia** has limited indigenous natural gas reserves, but its consumption exceeds its supply and has been a net natural gas importer since 1983.[88] It imports its needs via pipeline from Algeria, that continues to Italy. Thus, despite being a neighbouring country to the region, Tunisia has a minimal

---

[87] https://www.gem.wiki/Delimara_FSRU
[88] https://www.ceicdata.com/en/indicator/tunisia/natural-gas-imports

influence on the Eastern Mediterranean natural gas market dynamics.

- **Albania** has an EEZ dispute with Greece. It has limited indigenous natural gas reserves and currently, its production suffices its demand. Thus, despite being a neighbouring country to the region, other than being a route for one important pipeline, Albania has little influence on the Eastern Mediterranean natural gas market dynamics.

- **Spain** is the EU's fifth largest natural gas importer and, as an EU member, its natural gas policies are governed by EU overall policy. Its natural gas imports come partially via pipelines from Algeria and France, (despite French obstacles as seen in the case of MidCat pipeline discussed earlier, see Section 4.2),[89] although the majority of its imports are supplied via seven operating LNG regasification terminals. Spanish investors were partners in Egypt's Damietta LNG liquefaction terminal, and they won compensation from Egypt when these terminals were idle. They divested of their investments in 2021. Spain historically imported LNG from Egypt and resumed this in 2022. It has the potential to import more natural gas from the Eastern Mediterranean Region, especially if more liquefaction capacity is established.

- **United Kingdom** used to possess significant indigenous natural gas reserves and production. Currently, the majority of these are depleted and, at the moment, its needs exceed its production, so the country is a net natural gas importer. It imports its natural gas via pipelines from Norway and Belgium, plus via three LNG regasification terminals. The UK historically imported LNG from Egypt and, after an interruption, resumed this in 2022.

---

[89] https://www.statista.com/statistics/1230118/natural-gas-imports-by-country-of-origin-spain/#:~:text=Liquefied%20natural%20gas%20imports,-Home%20to%20the&text=In%20 addition%20to%20the%20energy,alternative%20to%20Russian%20gas%20supply.

As mentioned earlier, if the UK were ever to consider pursuing claims for the EEZs of its sovereign bases in Cyprus, it would have EEZ disputes with Cyprus, potentially also with Turkey, Egypt, Israel and Palestine. This would considerably complicate the geopolitical scene in the Eastern Mediterranean, particularly since the country is a founding member of NATO and is a permanent member of the UN Security Council.

- **UAE** has the significant crude oil reserves and ranks sixth in the world. It is the world's eighth largest crude oil producer, and ranks sixth net exporter. It is a member of the Arab League, OPEC and OAPEC. The country recognises Israel.

In terms of natural gas UAE has the world's seventh largest reserves of natural gas,[90] and is the world's fifteenth largest producer. However, currently it is a net importer as its consumption exceeds its production. Its imports come via pipeline from Qatar and via two LNG regasification terminals in Dubai and Abu Dhabi. Despite this it exports LNG from its ADNOC LNG liquefaction terminal at Das Island, and has started constructing a new terminal at Ruwais. The country aims to increase its oil and natural gas production and has started developing several reserves both conventional and unconventional. The country is a member of GECF.

From natural gas perspective, at the time of writing, the UAE has little influence quantitatively on the Eastern Mediterranean natural gas market dynamics, although one of its companies (Mubadala) acquired a stake in Tamar, one of Israel's main natural gas fields,[91] and its national oil company ADNOC is attempting to acquire a stake in the Israeli company NewMed, which owns substantial

---

[90] Reserves reported as 289.928 Tcf at the end of 2023; reference: https://www.ogj.com/general-interest/economics-markets/article/14302481/global-oil-and-natural-gas-reserves-both-increase
[91] https://www.ft.com/content/0188209c-594a-45d6-959c-26a3568db7e1?desktop=true &segmentId=dd5c99e9-30be-ddd0-c634-ff3a0c2b738f#myft:%E2%80%A6

proportion in Israel's natural gas reserves. Also, Emirati companies hold stakes in several Egyptian and Cypriot gas production and exploration blocks. Natural gas volumes aside, politically the UAE is the chief manipulator in the region, and its influence on the events in the Eastern Mediterranean region and by default natural gas markets is significant.

- **Saudi Arabia** has the largest conventional crude oil reserves and ranks second behind Venezuela if heavy oil is included. It is the world's second[92] largest crude oil producer, and largest net exporter. It is a founding member of the Arab League, and a founding member and chief influencer of OPEC, as well as a founding member of OAPEC. To date, Saudi Arabia still does not recognise Israel.

In terms of natural gas Saudi Arabia has the world's sixth largest reserves of natural gas,[93] and is the world's ninth largest producer. However, currently it is self-sufficient and consumes all its production. It aims to increase its production to use in electricity generation freeing additional crude oil for exports. It has started developing a giant unconventional field Jafurah in 2021,[94] which has 229 Tcf reserves.[95] Saudi Arabia is not a member of GECF.

If Saudi Arabia decides to become net natural gas exporter it can choose to export via LNG liquefaction terminals or pipelines. Sending natural gas by pipelines can link to Turkey via Jordan and Syria to supply Europe, or potentially via Jordan and Israel to link to Egypt to supply Europe. The latter option is politically suicidal for now especially with Gaza war still raging. But never say never.

---

[92] Alternates with Russia, influenced recently by OPEC+ strategy.
[93] Reserves reported as 335.997 Tcf at the end of 2023; reference: https://www.ogj.com/general-interest/economics-markets/article/14302481/global-oil-and-natural-gas-reserves-both-increase
[94] https://www.offshore-technology.com/projects/jafurah-gas-field-saudi-arabia/#:~:text=The%20field%20is%20owned%20and,EPC)%20contracts%20worth%20%2410bn
[95] https://www.aramco.com/en/news-media/news/2024/aramco-adds-significant-volumes-to-proven-gas-and-condensate-reserves-at-jafurah

- **Azerbaijan** has significant natural gas reserves,[96] and is currently a major natural gas exporter to Europe via SGC initiative. The country is an observer at GECF. From natural gas perspective, currently the country has little influence on the Eastern Mediterranean natural gas market dynamics but a major influence beyond that in the EU. However, it can be seen as a potential competitor or rival to East Mediterranean natural gas in supplying the EU and Turkey in the medium term.

- **Australia** has significant natural gas reserves,[97] and has transformed itself into a major LNG producer and exporter, competing with Qatar and USA. It started exporting LNG in 1989, but since 2004 it embarked on a massive cycle of establishing LNG liquefaction terminals totalling 87.8 mtpa,[98] which now dominate the Asian market, competing directly with Qatar, and thus they reduce the chance of any LNG imports from the Eastern Mediterranean penetrating the market. The appetite for new LNG projects has waned considerably and very few new projects are proposed.

- **China** has the world's eighth natural gas reserves, and is the world's fourth largest producer.[99] But despite this, it is the world's third largest natural gas consumer, and it's the third largest LNG importer. Following termination of its zero-Covid policy, its natural gas imports rebounded. As one of many sources, China historically imported. LNG from Egypt, and has resumed this in 2022. It has the potential to import more natural gas from the Eastern Mediterranean region especially if more liquefaction capacity is established. However, its impact on the Eastern

---

[96] Reserves reported as 60 Tcf at the end of 2023; reference: https://www.ogj.com/general-interest/economics-markets/article/14302481/global-oil-and-natural-gas-reserves-both-increase
[97] Reserves reported as 96.48 Tcf at the end of 2023; reference: https://www.ogj.com/general-interest/economics-markets/article/14302481/global-oil-and-natural-gas-reserves-both-increase
[98] Additional 5.4 mtpa are under construction.
[99] Reserves reported as 264.714 Tcf at the end of 2023; reference: https://www.ogj.com/general-interest/economics-markets/article/14302481/global-oil-and-natural-gas-reserves-both-increase

Mediterranean natural gas dynamics is minimal from a trade point of view, but its political meddling in the region cannot be discarded.

Chapter 5
IS IT A GAME CHANGER?

After discussing the situation of natural gas in each of the Eastern Mediterranean countries in detail, as well as the influence on the market from various other important countries, let us explore the value proposition, or the position of Eastern Mediterranean natural gas in the global context. Is it a game changer? Is it a mere footnote? Or is it somewhere in between?

To answer these questions, I needed to look at the bigger picture so as to give insights on the true situation. The following pages address this and look at the issue from different aspects.

Firstly, I quantitatively assess the volumes and place them in perspective to the global picture. Secondly, I discuss the power dynamics in the region to identify the winners and losers. At the time of writing, there is a real possibility that Donald Trump may regain the US presidency, and we all know he looks at the world in these terms i.e. loving winners and hating losers. Thirdly, I briefly touch on the elephant in the room. Will natural gas issues be a trigger for a war in the region? Finally, I look at the picture through a different lens, but one which is familiar to most people - using the analogy of the football game (that's soccer to the Americans) I will simplify and make the analysis clearer to the non-expert readers.

5.1 Putting the volumes in context

As already shown in Section 1.2, the world is awash with natural gas reserves, both conventional and unconventional. It is now widely accepted, by all major data sources, that reserves are growing - with additional volumes discovered, improved exploration technologies and better recovery processes from existing resource basins. Furthermore, as already discussed, significant volumes are now considered as reserves from unconventional shale gas, tight gas and coalbed methane plays. Production from these plays has altered the natural gas market globally, transforming the US from natural gas importer into a major natural gas exporter.

Thus, what is the big deal with Eastern Mediterranean natural gas reserves? Are they really a game changer? In a single word, the answer to this question is a resounding NO. All Eastern Mediterranean countries' natural gas reserves volumes summed up together, are still small, when they are considered in the global context.

As already discussed in Section 2.3, according to the American Geological survey and Germany's BGR, substantial resources, exceeding 340 Tcf of natural gas, are estimated to exist in the two largest Eastern Mediterranean basins. Therefore, even if all this resource volume is translated to reserves (see Appendix II to understand the concept of reserves versus resources), this only constitutes around 4% of global proven reserves.

Figure 5.1 below shows the reserves numbers for the ten analysed Eastern Mediterranean countries, as reported in Chapter 3. Figure 5.2 presents the combined reserves in the region, compared to the reserves of natural gas in selected countries. Note, the combined reserves of the region, if Libya is included, account for 2.6% of the global natural gas reserves reported by the Oil & Gas Journal at the end of 2023, but only 1.9% if it is excluded.[1] For sake of comparison, reserves in Qatar are 11.2% of global reserves – over four times more than combined Eastern Mediterranean reserves. They are even more in Iran, 16% of global reserves and – over sixfold relative to Eastern Mediterranean reserves. Furthermore, just consider this fact. According to the EIA, the USA added 152.1 Tcf to its reserves between 2020 and 2021, with a further 65.7 Tcf by end of 2022 – more than all the Eastern Mediterranean reserves

---

[1] Estimated reserves in the region not reported by Oil & Gas Journal are added to both combined regional data, and overall global data.

put together.[2,3]

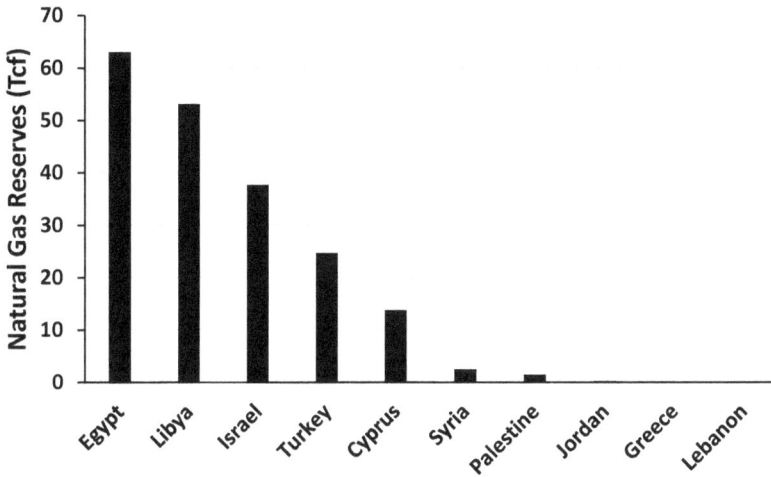

Figure 5.1: Summary of Eastern Mediterranean natural gas reserves (2023)

Sources: Oil & Gas Journal; Estimates by the author

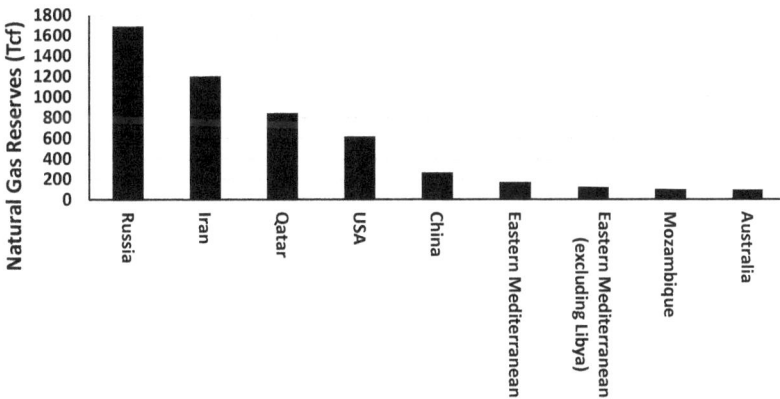

Figure 5.2: Comparison of natural gas reserves between the Eastern Mediterranean and selected countries (2023)

Sources: Oil & Gas Journal; Estimates by the author

---

[2] https://www.eia.gov/todayinenergy/detail.php?id=55339#:~:text=Proved%20reserves%20of%20natural%20gas%20in%20the%20United%20States%20grew,%2C%20Year%2DEnd%202021%20report.

[3] https://www.eia.gov/naturalgas/crudeoilreserves/index.php

Although the graphs above show reserves data, note that throughout this book, I have deliberately not presented historical natural gas (or oil) reserves data on a country level, since the reliability of these numbers is questionable and many believe that these numbers were inflated for political reasons.[4] In addition, the definition of what is considered reserves or resources differs country by country, depending on the economic feasibility of oil and natural gas extraction (see Appendix II for detailed technical discussion). Also, major data sources, such as BP/EI Statistical Review or EIA, do not publish up-to-date reserves data.[5] For example, it took several years to report any number for natural gas reserves for Mozambique, despite fields developments being underway. The delay in publishing up-to-date reserves data is similar with Tanzania, Senegal, Mauritania, Suriname, and Cyprus. Similarly, oil reserves in Guyana took years to make it into these data sources.

If we look at the role of Eastern Mediterranean natural gas in the global context, in terms of production, the numbers are, to put it mildly, small. Figures 5.3 and 5.4 show the numbers from the ten Eastern Mediterranean countries and benchmark them against selected countries, such as Iran, Qatar, and Australia. As can be seen, the numbers illustrate the minor role they play in the global market.

However, with most production consumed locally (5.1 Tcf a year in 2023), volumes available for export are even smaller than production numbers. At the moment only Israel (0.5 Tcf/y) and Libya (0.2 Tcf/y) are net exporters and, even when they are combined, only account for around 1.5% of total export markets. In this instance, I did not include Egypt's LNG exports in this number, as this would constitute double counting, as Egypt re-exports surplus volumes imported from Israel, since it became net importer.

---

[4] See a detailed discussion in my previous book "Fossil Fuels in the Arab World – Facts and Fiction".
[5] According to these data sources, they claim that they report these numbers with a time delay when they are certified as reserves only.

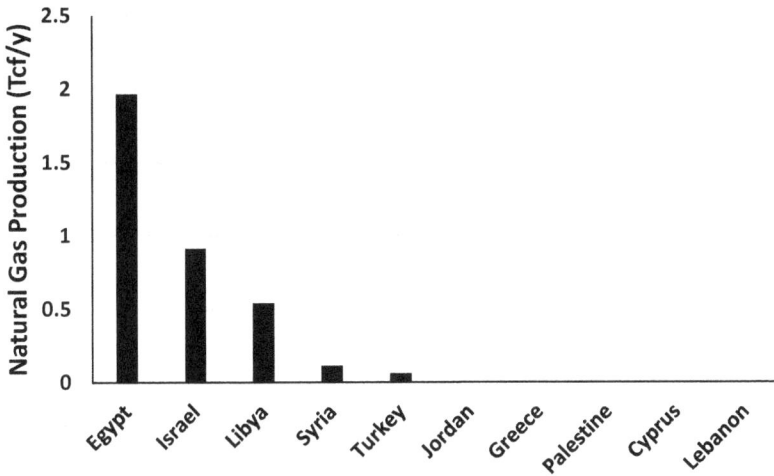

Figure 5.3: Summary of Eastern Mediterranean natural gas production (2023)

Sources: Oil & Gas Journal; Estimates by the author

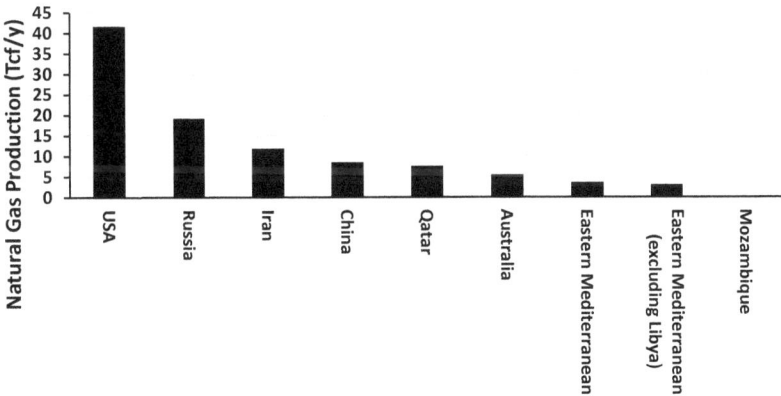

Figure 5.4: Comparison of natural gas production between the Eastern Mediterranean and selected countries (2023)

Sources: Oil & Gas Journal; Estimates by the author
Note: Mozambique production has grown substantially in 2024.

The majority of the natural gas trade in the region is internal. Libya exports to Italy via pipeline, as its LNG exports stopped over a decade ago. Most of Israel's exports are intra-regional to Jordan or to Egypt for domestic consumption. Prior to Israel's deal with the

EU (to export natural gas via Egypt), only excess Israeli natural gas exported to Egypt but not needed internally in Egypt, was exported and also, diminishing amounts of Egyptian natural gas were also exported. Thus, if we look at the exports to outside the region, rather than intra-regional trade, the numbers shrink further.

With more natural gas reserves yet to be found and additional production planned from yet to be developed reservoirs, there is imminent potential for additional export volumes from Israel, Cyprus, Libya, with, later down the line, export volumes from Lebanon and Syria. Obviously exact numbers for this potential are still speculative and pure guesses, despite numerous wild claims in the media.

While the majority of net exports from the region are via pipelines, it is expected that the future exports will be in the form of LNG, either using Egyptian infrastructure or proposed infrastructure in Israel and Cyprus. In the future, pipeline projects will take a backseat, and many proposed pipelines will remain pipedreams. One main example is the EastMed pipeline.[6] By their nature, transport pipelines are vulnerable structures – many of them, even deep offshore pipelines can be attacked, as proved by the attacks on the Russian Nord Stream pipeline in 2022. As energy transition (from fossil based to 'Green' energy) is creating uncertainties about the medium to long-term demand for natural gas, thus threatening projects with high capex investments that need long payback time, the global trade is moving into LNG which offers more flexibility.

However, with all of the political disruption and instability in the Eastern Mediterranean Region, potential LNG projects in Israel and Cyprus face an uncertain future with continued delays preventing progress. At the same time, Qatar and the USA are forging ahead with their LNG projects increasing capacity. So, time is of essence and further delays in Israel and Cyprus risk both countries missing out on their moment of opportunity.[7]

---

[6] https://www.naturalgasworld.com/eastmed-the-beginning-of-the-end-gas-in-transition-96699
[7] https://www.ft.com/content/0188209c-594a-45d6-959c-26a3568db7e1?desktop=true
&segmentId=dd5c99e9-30be-ddd0-c634-ff3a0c2b738f#myft:%E2%80%A6

Overall, total net exports are around 1.5% of the global market and LNG exports are around 0.9% of the LNG global market, plus the LNG exports are unreliable with many suspensions, interruptions and so, cannot be depended upon.

In fact, overall natural gas production and consumption numbers analysis shows that the Eastern Mediterranean Region is currently a net natural gas importer, with Turkey and Egypt responsible for the majority of consumption. This is illustrated in Figure 5.5.

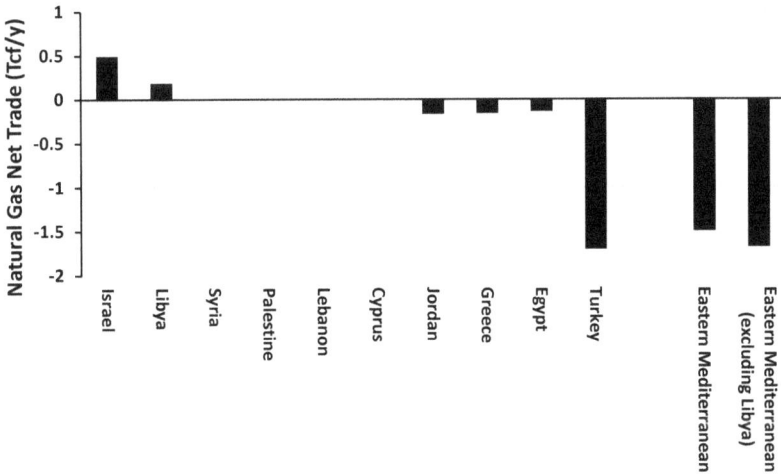

Figure 5.5: Summary of Eastern Mediterranean natural gas net trade (2023)

Sources: Oil & Gas Journal; Estimates by the author

So, to summarise, the Eastern Mediterranean Region's importance to global trade is due to its location. The location is strategic because of its vital trade routes i.e. Suez Canal and various pipelines, but the volumes of its natural gas are not so important. In reality, the region is net importer, despite the hype by local media regarding its natural gas export potential. Its natural gas exports effect on market are trivial, they hardly register on global trade, as most trade is intra-region and even there, the overall trade is still small.

Regionally, as it stands the balance of influence has tilted towards

Israel. Egypt's potential is fading in parallel with its declining production. Although rapid development in Cyprus or Libya could transform the balance of influence by producers in the region, this seems unlikely in the short term. However, if these developments were to materialise, or if significant new reserves were discovered in Egypt, Syria and Lebanon for example, the balance of power could tilt again, perhaps favouring Egypt or giving clout to Syria. Indeed, there are several scenarios that exist, and each might change the power play in the future.

The slow progress or inaction by many countries in the region benefits other countries that are competing to provide the natural gas to the region's export targets. These countries may even supply natural gas to parts of the region if the region fails to realise its own indigenous potential.

Finally, to put a potential monetary worth of all reserves and exports in US$, a simple calculation is conducted assuming a natural gas price of 10 US$/MMBtu. The numbers come up to around 2040 billion US$ for the entire reserves and around 700 million US$ per year (based on 2023 numbers) in exports. Obviously, these theoretical numbers do not equate to net profits as a large proportion of these values will be spent as expenses.

There are several myths regularly published in the media, that I mentioned earlier. The following three are the major ones that need to be addressed:

1. That the natural gas discoveries in the Eastern Mediterranean are potentially massive and will be significant to the world energy balance. As demonstrated earlier, this is incorrect because the volumes are in fact small, and it will face tough competition finding a market outside the region.

2. That there are lots of additional reserves of oil and natural gas in the region, i.e. Syria or Greece, which are either not found yet, or hidden, due to political reasons. Again, this is incorrect, as scientific studies those carried out by USGS

or BGR, estimate low volumes, as opposed to optimistic or deliberately exaggerated numbers. As discussed earlier, there is no evidence of Kuwait size reserves in Syria,[8] nor that Greece has 27 trillion m^3 of natural gas.[9]

3. That the Syrian civil war was mainly over a natural gas pipeline from Qatar to Syria. The fate of the pipeline depends on who reports the story. This is a total myth as discussed earlier.[10]

All these myths and more are rife in the Arab world. And following the 2023-present Gaza War, their prominence has amplified several folds.

5.2   Winners and Losers

Several countries in the Eastern Mediterranean Region are competing to establish themselves as energy hubs. Each of them is working hard to enhance its position, publicise its credentials and, in some cases, even sabotage its competitors' efforts. I summarise below where the players stand.

I have created the diagram below to illustrate the complexity of the multi-dimensional relationships between all of these countries. As you can see, with just a glance, how many links, overlaps, parallels and duplication there are between the parties involved. Figure 5.6 below shows the natural gas flow between all the stakeholders. It resembles a map of the resources and competing pipelines, many of which have led to conflicts. The diagram shows Libya and Israel, the two countries in the Eastern Mediterranean Region exporting natural gas internationally: the former via its pipeline connection to Italy and the latter, which uses Egypt's LNG export facilities. The majority of natural gas production in the Eastern Mediterranean is consumed within the region's borders. Significant volumes of natural gas that are travelling through the

---

[8] https://www.youtube.com/watch?v=0UvCUPCBsgc
[9] https://slpress.gr/english-edition/an-untold-treasure-await-extraction-from-the-greek-seabed/
[10] https://www.middleeasteye.net/big-story/pipelineistan-conspiracy-war-syria-has-never-been-about-gas

region, are actually shipments of natural gas from distant producing regions in Russia and the Caspian, which will potentially travel further afield to countries in Europe.

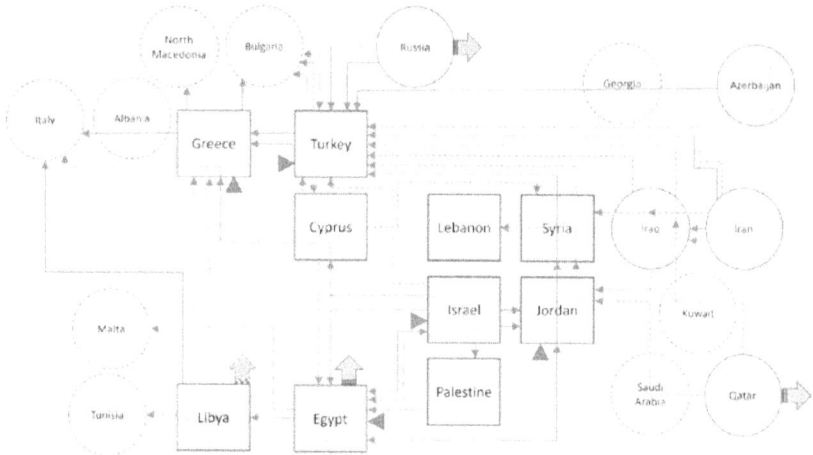

Figure 5.6: Natural gas flow within the Eastern Mediterranean region and beyond its borders.

Source: by the author
Note 1: The connections do not include cancelled or abandoned projects.

As discussed in Chapters 3 and 4, the main three countries competing to become the main energy hubs in the region, both as a way of progressing their economies and to exert more influence in the region, are Egypt, Turkey and Greece. As natural gas is now being treated as an essential part of the energy transition pathways, these three countries are pushing these natural gas credentials to gain every advantage.[11]

Italy and Spain are also part of the energy trade equation. However, despite both countries possessing significant LNG regasification import capacity, as well as both being connected to Algeria and Libya with natural gas pipelines, their efforts to increase their supplies of natural gas are hindered by stubborn persistent obstacles. The most significant of these are caused by France, which for political reasons, continues to resist establishing better

---

[11] Even Malta wanted to be a hub – see https://emialliance.com/malta-eager-to-become-the-new-energy-hub-of-the-mediterranean/

connections with these two EU neighbours, and the other most significant one is the sporadic breakdowns in diplomatic relations between Algeria and Morocco, affecting current and proposed export routes.

Other countries in the region tried to compete in this field, but they have decided to change strategy e.g. Jordan which is now focusing on strengthening its energy power links with Iraq, Egypt, Palestine, Israel and Saudi Arabia, to become an electricity corridor, as discussed before.

One of these top three players, Egypt, is capitalising on its geographical location and existing infrastructure in order to establish itself as THE central oil, natural gas, electricity, and, recently, renewables hub. However, although its location and infrastructure are absolutely pluses, it has significant weaknesses. Firstly, it has returned to being a natural gas net importer, after years of being a net oil importer. As a result of this, its provision of electricity to its customers is interrupted daily which means that it is in absolutely no position to attempt to become an electricity hub. Egypt's plans to attract investment in green energy have stalled because of the dire economic conditions in the country. Secondly, since the government is weak and cannot prevent the army interfering in all economic sectors, there has been a deterioration in the economic conditions, exacerbated by the endemic corruption which shows no signs of reversing. These conditions raise the odds against competitive success in the market considerably. Thirdly, the geopolitical instability of its neighbours (Libya, Sudan, Israel and Palestine) has spilled its borders, threatening its hopes to be a safe trade route or a trusted supply chain logistics hub.

Although Egypt is a de-facto natural gas hub, with Israel using its natural gas infrastructure to export LNG and Cyprus planning to follow suit, its position has been reduced from a major exporter to a mere service provider that collects a nominal fee. Both Israel and Cyprus are already examining alternative options to working with this problematic business partner. Egypt's plans of building alternative export natural gas pipelines indefinitely postponed, in fact, all the pipelines linked to Egypt are for importing gas only.

One of the 'top three' contenders, Greece, has different ambitions. It has always been a net energy importer. Its strategy relies on successfully establishing itself as an energy gateway to Europe. Initially this strategy made headway, as it had several natural gas pipelines in operation, but it has a major weakness, which is the reliance on Turkey. Turkey is long-standing adversary of Greece with whom it has a long history of territorial disputes. Turkey controls the main supply route for natural gas from the Caspian and potential future flows from Central Asia or the Middle East. Meanwhile, although an expansion of import capacity is planned, Greek reliance on LNG regasification terminals as an alternative has shown only limited success.

The Greek strategy of promoting the EastMed pipeline as the cornerstone of natural gas integration in the region, but excluding Turkey, failed spectacularly. All attempts to resuscitate the project in either its current form or via Egypt or Libya are doomed to fail.

Turkey, out of the three countries, is showing the most potential and promise. The country has employed a long-term game, and it is already an energy hub. Numerous oil and natural gas pipelines pass through the country, linking Russia, the Caspian, Central Asia, Iran and Iraq to global markets. In addition, there are several proposed projects on the drawing board. Despite the public disagreements and bad rhetoric between Turkey and many of its neighbours, the country has had great success in separating its economic/political agendas while continuing to maintain close economic cooperation with Iran and Israel.

To bolster its position, Russia publicly supports Turkey in its ambitions to turn itself into a natural gas hub. Also, the rapid development of the natural gas discoveries in Turkey's Black Sea will strengthen its position, as it can supply an important portion of its natural gas demand. This could also help ease some of the tensions in the Eastern Mediterranean if Turkey were to intensify its efforts to explore natural gas in the Black Sea, rather in disputed

Cypriot or Greek waters.

In recent years, Israel has transformed itself into the region's main gas exporter. Foreign investment is pouring into Israel's efforts to find and develop more natural gas projects. This can be seen by the presence of major IOCs in the country, as well as the involvement of UAE companies and possible involvement of Saudi companies in the future. Superficially, it might seem, at a glance, that Israel appears to be the prospective 'main player' in this natural gas game. However, Israel's aggressive foreign policies and military attacks against its neighbours may jeopardise all of its plans to develop as a global natural gas supplier. The 2023-present Gaza War is already having a negative impact on the country's image, and it is yet to be seen if this will affect future investments, as the country continues to be firmly supported by the West. Thus, future development in Israel is vulnerable and any lack of support from the US or Europe may influence the future trajectory of how Israel's resources are developed. Of course, this will have knock-on effects on resources development in the whole region.

One important obvious fact though, the current war has derailed, yet again, any natural gas development in the Palestinian territories.

Libya continues to be an important oil and natural gas exporter. However, its potential has been sabotaged by the political instability in the country. The situation is even worse in Syria, as the political and security turmoil is preventing the country from even considering pursuing meaningful exploration or development in the foreseeable future.

Finally, despite the years of confirming significant reserves, numerous proposals for natural gas pipelines and LNG terminals, sadly the disagreements Cyprus has between its stakeholders has prevented any development. For the time being, Cyprus appears to be stuck, in terms of natural gas development.

As for Lebanon, despite all its hopes and dreams, no commercial reserves have been found yet. In addition, the Israel-Gaza 2023-present war and Hezbollah's involvement in the conflict reduces the countries chances further.

As discussed earlier, from a natural gas market dynamic, the situation is very polarised as 'against-Turkey' versus 'with-Turkey'. As it stands, there is a winner from each camp, while I leave it to the readers to assess who are the losers.

5.3   War or no war?

Many erroneously continue to predict that the next major war in the Eastern Mediterranean will be an energy war. In fact, it will most probably be a war over water resources. Energy in the region is abundant, fossil fuels can be substituted with renewables. However, water is scarce. Dams are being built along major rivers and diverted by the following countries- Israel, Turkey, Ethiopia and Iran, which seriously threatens life in downstream countries such as Jordan, Syria, Iraq and Egypt.

Moreover, at the time of writing a major war is being fought in the region, the 2023-present Gaza War is yet to conclude. Despite what some believe, it is not an energy war. It is a war over land and ideology.

There almost appears to be an 'oil curse', where those countries with rich oil deposits in the region seem to be plagued by war. This is happening in Syria, Libya, Yemen and Sudan. The reality is that, in all these countries, oil was not the reason for the wars, but its existence certainly helped fuelling them.

5.4   The football game analogy

What is happening in the region is a geopolitical game that uses natural gas policies as an important strategy. This 'game' comparison can be further simplified for the non-experts by using some sort of analogy. We could consider the game as a theatrical play, where the first act is well under way but the characters are playing their roles without rehearsal or perhaps as a football (soccer to the Americans) match which is descending into chaos, the players have lost all discipline and there is no referee to ensure fair-play or issue players who commit fouls with yellow or red cards.

In this section I have chosen to present the situation using that analogy of a football match, with football teams and players. I know this maybe an odd approach when writing about natural gas, but I think this may bring the subject closer to the non-experts, some of whom may believe all the conspiracy theories but also, many of them believe that they are football experts. Therefore, since not many people know natural gas terminology, but many more know football, by writing this section, I am trying to reach a wider audience to give a clearer picture, explain the situation, provide the historical context, deliver the message and propose some ideas. Also, these countries have keen football fans with great teams. FIFA and UEFA are sponsored, in part, by fossil fuel money so there is already a strong link between football and natural gas.

Recall if you will, the 2022 FIFA World Cup in Qatar. Natural gas revenues generated by QatarEnergy, which is the company behind that World Cup host's wealth, were instrumental in enabling the hosting of this tournament and making it a great success. Since the country was awarded the tournament in 2010, it has poured, at least, an eye-watering 200 billion US$ into new infrastructure, including 6.5 billion US$ on stadiums and facilities, largely paid for with revenues from QatarEnergy's LNG exports.

First, as we would with a soap opera, let's allocate each country a character or role from the football world, and let's select famous men from the oldest, most popular league in the world, English Premiership (both past and present).

- Egypt is the once great but now fading star, past his prime, but can perform brilliantly on occasion. Think of Romelu Lukaku, or Eden Hazard.

- Israel is the promising talent, young, maturing well, but lacking the essential traits that could transform into a top star. Think of Mason Mount or Callum Hudson-Odoi.

- Syria is an experience player but who got an injury before the season started and is now 'on the bench' indefinitely. Think of Jonathan Woodgate or Reece James.

- Libya is an average team-player, one with potential but unluckily prone to injury. Think of Andy Carroll or Arjen Robben.

- Greece is the frustrating also-ran player; he still thinks of himself a star but often plays below par and underperforms. Think of Michy Batshuayi or Danny Drinkwater.

- Turkey is the rogue player, talented but uncontrollable, prone to getting red cards and being sent off the pitch. Think of Joey Barton, Mario Balotelli or Diego Costa.

- Cyprus and Lebanon are below average players. The ones who miss the pass, easily loose the ball when challenged, do not contribute much and miss the penalty shoot-out opportunities at the end of a hard-fought match. For Lebanon, think of Dele Alli and for Cyprus, picture Harry Maguire, they have their good moments, but leave the fans with their heads in their hands.

- Jordan is the talented player with occasional brilliance, who tries hard but is often unlucky, often misfires and misses the mark. Think of Timo Werner or Fernando Torres.

- Palestine, a player completely out of their league, a one-legged para-athlete, outmatched on every level, an amateur playing amongst professionals, with no chance in the game.

Let's get acquainted with the main players, look at their

performance, consider how each of them may develop in the future, who has potential and who has passed their prime. As with many elite athletes who are aware of their financial value and have many titles or awards to their name, we can see that each of these players think mostly of themselves, playing to bolster their own reputation or interests rather than playing for the whole team.

However, the global natural gas powers in the world gas market, when it comes to global natural gas contest, the players in this fantasy Eastern Mediterranean League, although they imagine themselves to be playing at the most elite level, actually play in the lower tiers, the equivalent of League Two and not in the Premiership.

Let's expand the analogy to include the supporting players. Here, I am using the analogy of the games structure, rules, governing bodies and authorities, its officials and influences rather than players on the pitch, as this can demonstrate what is happening more clearly. Note this is done with reference to the Eastern Mediterranean League. In this case, a significant character can be designated as follows:

- USA plays the role of a FIFA (Federation International de Football Association) official, representing a larger power that oversees the game, i.e. the global contest. It is often in disagreement with Russia. It is prone to corruption and backdoor deals.

- Russia plays the role of a UEFA (Union of European Football Associations) official, equally important to FIFA in footballing terms as it has similar resources. It is in disagreement with FIFA, but is also rife with corruption and behind closed doors deals.

- France plays the role of an FA or Premiership official, often interfering, trying to impose its will in the regional context.

- Iran and Iraq play the role of senior and junior football

investors, who partnered and have big vision, grand plans but, for various reasons, they are poorly executed, leading to their strategy ending in disappointment.

- Qatar plays the role of Sky TV, massively influences by offering good expensive broadcasting packages, that affect all of the football world.

- Saudi Arabia's role is of an international tv right holder, that has some influence but plays second fiddle to Sky TV.

- UAE and United Kingdom are VAR's (video assistant referees) officials, who influence the game from the sidelines but from a somewhat unreliable position.

- Italy and Algeria roles resemble football season ticket holders, always connected to the game, but do not influence it.

- Albania, Malta and Tunisia are ordinary spectators.

- Spain and China are VIP spectators.

- Azerbaijan and Australia's roles are of TV pundits, who comment and shape opinions from afar. They have some influence affecting the overall season sentiment, but not a specific game or the immediate outcome.

So, with all roles defined, the tournament is underway, with all the players, officials and others ready for the games to begin. It will be interesting to watch how it plays out. Earlier, I asked, 'Is it a game changer?' and the answer is clear to me, may be to most of the readers here. This is a game being played in a lower division. League Two, of course, it is important to the players in that League, but it has no effect on who will win the Premiership.

In my opinion, as it stands now, this is a football match gone awry. It is filled with competing egotistical players, acting as individuals rather than playing as a team. The game is being played without a referee and the result is chaos. While that chaos continues, external interests can come in, hoover up the best talent, use the resources and profit from the income generated while the source of the talent, the facilities for developing that talent are left to languish – always competing, never winning the ultimate prize. This analogy reflects the disordered geopolitical landscape in the Eastern Mediterranean Region, where everyone is focused on their own interests rather than working towards a collective goal.

I trust that the message can be delivered this way, communicated and understood clearly by all. When we look at a system that is failing itself and failing most of those it should be serving, when we can clearly identify where it is not working and why – therein lies the seeds for correcting it and transforming it into the best that it can be. While the Eastern Mediterranean natural gas resources may never be enough to be considered significant on a global scale, still, with the right conditions, it can make a difference to the region, its population, the climate and the energy picture in its part of the world. The Arabic proverb 'رُبَّ ضارَّةٍ نافعة', which translates to 'Sometimes harm can be beneficial', reflects the idea that challenges and apparent failures can contain the seeds for improvement and success, much like how the identified flaws in the system or the Eastern Mediterranean's modest natural gas resources can still hold potential for significant positive change under the right conditions that benefit the whole region and spread beyond its borders.

APPENDICES

Appendix I: East Mediterranean Countries' Statistics

Table A1.1: East Mediterranean Countries' basic information

Country	Capital	Area km²	Rank	Population million	Rank
Cyprus	Nicosia	9251	9	1.703	10
Egypt	Cairo	1001450	2	111.247	1
Greece	Athens	131957	5	10.461	5
Israel	Jerusalem	21937	7	9.403	6
Jordan	Amman	89342	6	11.174	4
Lebanon	Beirut	10400	8	5.364	9
Libya	Tripoli	1759540	1	7.361	7
Palestine	East Jerusalem	6220	10	5.385	8
Syria	Damascus	187437	4	23.865	3
Turkey	Ankara	783562	3	84.120	2
East Mediterranean Region		3999801		270.084	
Region's Share globally (%)		2.69		3.35	

Source: CIA Factbook, 2024.
Note 1: Population as estimated in 2024.
Note 2: Area and population of Syria include occupied Golan Heights and its Arab population only.
Note 3: Area and population of Palestine is for the West Bank including East Jerusalem and Gaza Strip. Population number excludes Israeli settlers.
Note 4: East Jerusalem is the proclaimed Palestinian capital. Ramallah is the current administrative centre of the Palestinian Authority, while Gaza city is the administrative centre of the Hamas-led government.
Note 5: Jerusalem is the proclaimed Israeli capital. Tel Aviv-Yafo hosts most diplomatic missions.
Note 6: Area and population of Cyprus include the self-proclaimed Turkish Republic of Northern Cyprus (TRNC).

Appendix II: Fundamental Fossil Fuels Terminology: Understanding the Terms 'Resources' and 'Reserves'

Here, I revisit an important subject from my previous three books.[1,2,3] I believe is essential to provide a thorough understanding of this area so that the events and consequences discussed in this book are as clear as possible to my readers.

The media has a habit of making incorrect, loose usage of terminology and definitions. This deliberate, ignorant, intentional or unintentional, use of a mix of technical engineering and geological terms, often causes misinformation, leading to inaccurate conclusions.

While the text in this section is not a dictionary of all these terms, it will provide clarity about the two main terms most often mixed up when discussing or reporting oil and gas issues, i.e. 'reserves' and 'resources'.

The definitions of fossil fuel resources and reserves are summarised in this section. If you do not already have a copy in your library, for a more comprehensive discussion, please refer to my previous books.

In the most simplistic terms, *fossil fuel resources* are defined as the fossil fuels initially-in-place, i.e. the quantity of fossil fuels, in varied sized deposits and reservoirs, that exist originally on, or under the earth's crust, in naturally occurring accumulations. These resources 'initially in place' are referred to by numerous names including *resource base, hydrocarbon endowment* or *original-in-place*.

Two main criteria categorise fossil fuels resources: discovery and

---

[1] Basel Nashat Asmar, Fossil Fuels in the Arab World: Facts and Fiction – Global and Arab Insights of Oil, Natural Gas & Coal, 2050 Consulting, London, UK, 2010

[2] Basel Nashat Asmar, Fossil Fuels in the Arab World: Seasons Reversed – Oil and Politics Interplay in the Arab World, 2050 Consulting, London, UK, 2017

[3] Basel Nashat Asmar, Fossil Fuels in the Arab World: Missed the Boat? – Adjusting to the Post-Oil Era, 2050 Consulting, London, UK, 2022

recovery. Accordingly, in terms of discovery, fossil fuels are either discovered or undiscovered. In terms of recovery, they are either recoverable or unrecoverable.[4] Recoverability is assessed at present day technology and feasibility. It will vary with technical advances and price changes. The criteria to assess discovery, recovery and commerciality of resources are well defined but vary between standards. Different definitions apply across the financial regulations that govern various stock markets where companies are listed. Interested readers can consult specific standards for the details of applied criteria.

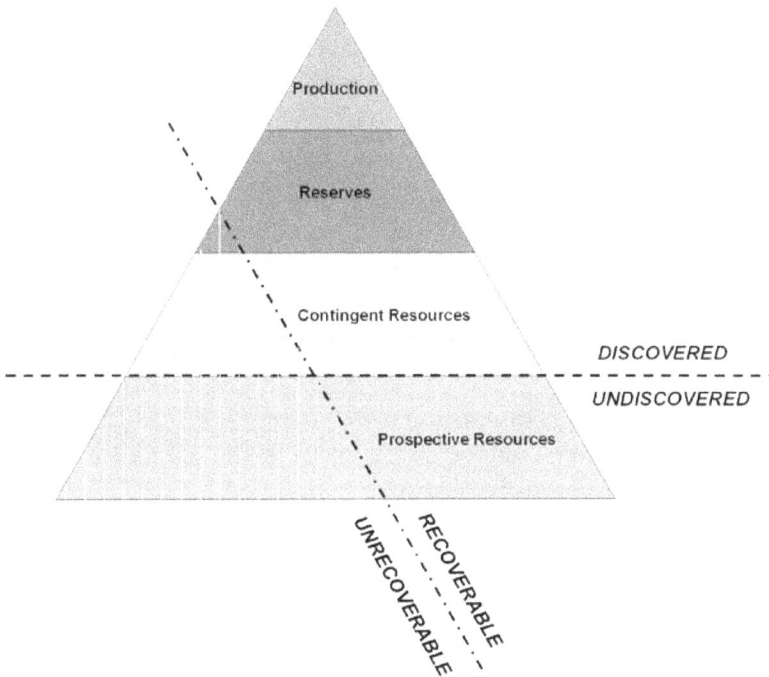

Production

Reserves

Contingent Resources

DISCOVERED

UNDISCOVERED

Prospective Resources

UNRECOVERABLE   RECOVERABLE

Figure A2.1: Categorisation of fossil fuels resources

Source: B.N. Asmar, Fossil Fuels in the Arab Worlds: Facts and Fiction, 2050 Consulting, 2010

The relationship between the above two criteria is illustrated in

---

[4] Refer to Figure 3.2. Unrecoverable category is relative. It becomes smaller without commercial constraints as one could probably extract more of the resources then.

Figure A2.1, which places the reserves within the framework of total resources. The figure also shows that three distinct portions of resources exist: *production*, *recoverable* resources and *unrecoverable* resources. It is worthwhile to note that the diagonal recoverable-unrecoverable interface can move both to the right, or the left, as a result of several interacting factors including fossil fuel price, extraction technology, political risks and legislative conditions, to name but a few.

In simplistic terms, the production refers to the cumulative quantities of fossil fuels already extracted, which are clearly recoverable. The remaining resources are either recoverable or unrecoverable. The sum of cumulative production and recoverable resources is often referred to as *ultimately recoverable resource* (URR) or *estimated ultimate recoverable* resource (EUR).

The remaining recoverable resources are divided into discovered and undiscovered.[5] In turn, the discovered resources are further divided into commercial and sub-commercial. Amalgamating these definitions together for simplicity, the recoverable resources can be described as divided into three distinct categories:

- **Reserves** - which are quantities of fossil fuels that are discovered, remaining in the ground awaiting production, recoverable and commercial. Reserves are further sub-categorised in accordance with the level of uncertainty associated with the potential recovery, due to both feasibility and technology, with any of the two factors rendering the reserves proved, probable or possible.
- **Contingent Resources** - which are quantities of fossil fuels that are discovered, remaining in the ground awaiting production, potentially recoverable however they are not yet commercial. Similar to reserves, contingent resources are further sub-categorised in accordance with the level of uncertainty associated with the potential

---

[5] These undiscovered resources are often referred to as yet-to-find (YTF).

recovery, due to both feasibility and technology, with any of the two factors rendering the reserves low, best or high estimates.

- **Prospective Resources** - which are quantities of fossil fuels that are undiscovered, remaining in the ground awaiting production, potentially recoverable and potentially commercial. Similar to reserves and contingent resources, prospective resources are further sub-categorised in accordance with the level of uncertainty associated with the potential recovery, due to both feasibility and technology, with any of the two factors rendering the reserves low, best or high estimates.

Figure A2.2 illustrates the above categories and sub-categories, where the dynamic, elastic nature of the boundaries defining the resources and reserves can be seen. Ultimately it illustrates the probability that a project will be developed and can reach commercial producing status. The figure presents the classification framework of fossil fuel resources, as defined in the SPE-PRMS,[6] where the horizontal axis represents the range of uncertainty, which is a degree of geologic assurance and recovery efficiency. This in turn, reflects the range of estimated quantities potentially recoverable from an accumulation and the vertical axis represents the chance of commerciality.

The unrecoverable resources are the portion of both discovered and undiscovered fossil fuels initially in place, which cannot be recovered at a given date due to geological, commercial accessibility or political constraints. However, a portion of these quantities may become recoverable in the future.

---

[6] The Society of Petroleum Engineers (SPE) Oil and Gas Reserves Committee, made up of international oil and natural gas experts, partners with several industry related societies to provide publicly available resources for the consistent definition and estimation of hydrocarbon resources. As part of this work, SPE offers free documents including the Petroleum Resource Management System (PRMS), the PRMS Application Guidelines, as well as a map to other systems, and standards for reserves estimating and auditing.

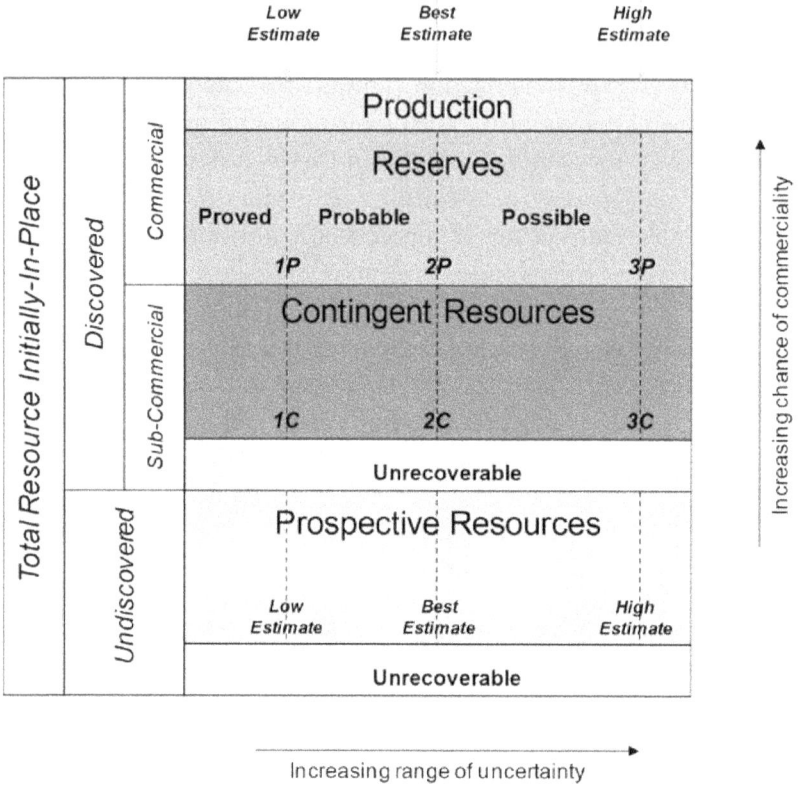

Figure A2.2: Classification framework of fossil fuel resources
Source: Modified from SPE-PRMS.

www.ingramcontent.com/pod-product-compliance
Lightning Source LLC
Chambersburg PA
CBHW061726270326
41928CB00011B/2133